STRATIGRAPHIC, PALEONTOLOGIC, AND PALEOENVIRONMENTAL ANALYSIS OF THE UPPER CRETACEOUS ROCKS OF CIMARRON COUNTY, NORTHWESTERN OKLAHOMA

The Geological Society of America, Inc.
Memoir 149

Stratigraphic, Paleontologic, and Paleoenvironmental Analysis of the Upper Cretaceous Rocks of Cimarron County, Northwestern Oklahoma

Erle G. Kauffman
U.S. National Museum
Washington, D.C. 20560

Donald E. Hattin
Indiana University
Bloomington, Indiana 47401

J. Dan Powell
Union Carbide Corporation
Grand Junction, Colorado 81501

1977

Library of Congress Catalog Card Number 76-47800
ISBN 0-8137-1149-5

Published by
THE GEOLOGICAL SOCIETY OF AMERICA, INC.
3300 Penrose Place
Boulder, Colorado 80301

Printed in the United States of America

Contents

Acknowledgments . . . x
Abstract . . . 1
Part 1. Stratigraphy and paleoenvironments . . . *Erle G. Kauffman, Donald E. Hattin, and J. Dan Powell* 5
Introduction . . . 5
Purpose and scope of the study . . . 5
Previous work . . . 7
Radiometry and zonal durations . . . 9
Descriptive stratigraphy . . . 11
Dakota Group . . . 11
Graneros Shale . . . 13
Lower shale member . . . 13
Thatcher Limestone Member . . . 13
Upper shale member . . . 14
Greenhorn Formation . . . 14
Lincoln Member . . . 14
Hartland Member . . . 15
Bridge Creek Limestone Member . . . 16
Regional stratigraphic relationships . . . 17
Comparison with Kansas section . . . 17
Comparison with southern Colorado sections . . . 19
Comparison with the Model anticline . . . 20
Comparison with south-central Colorado . . . 23
Comparison with north Texas section . . . 23
Summary of facies relationships of the Oklahoma section . . . 27
Biostratigraphy . . . 28
Western Interior . . . 29
Gulf Coast . . . 31
Correlation of the Cimarron County Cretaceous section . . . 31
Dakota Group . . . 31
Graneros Shale . . . 32
Greenhorn Formation . . . 33
Paleoenvironments . . . 38
Fossil assemblages . . . 39
Assemblage dominated by robust, byssate inoceramids and small ostreids . . . 39
Assemblage dominated by large, flat, thin-shelled inoceramids . . . 40
Assemblage dominated by robust, inequivalve, nonbyssate to slightly byssate inoceramids . . . 40
Assemblage dominated by small, thin-shelled pectinids . . . 41
Assemblage dominated by subequivalve, slightly to moderately biconvex, nonbyssate to weakly byssate inoceramids . . . 42
Offshore assemblage dominated by few ammonite taxa . . . 42
Offshore assemblage dominated by diverse ammonite taxa . . . 43
Paleoenvironmental interpretation: Physical and biological evidence . . . 43

Part 2. Paleontology . . . . *Erle G. Kauffman and J. Dan Powell* 47
Introduction 47
Systematic paleontology 49
Phylum Coelenterata 49
Class Anthozoa 49
Ahermatypic coral, indet. 49
Phylum Annelida 49
Class Chaetopoda 49
Family Serpulidae 49
Genus *Serpula* Linné, 1758 49
"*Serpula*" sp. 49
Phylum Brachiopoda 50
Class Inarticulata 50
Family Discinidae Gray, 1840 50
Genus *Discinisca* Dall, 1871 50
Discinisca? sp. 50
Phylum Mollusca 50
Class Bivalvia Linné (Buonanni, 1681) 50
Family Bakevelliidae King, 1850 50
Genus *Phelopteria* Stephenson, 1952 50
Phelopteria minuta Kauffman and Powell, n. sp. 50
Phelopteria quadrate species aff. *P. minuta* Kauffman and Powell 52
Phelopteria sp. A 52
Family Inoceramidae Giebel, 1852 53
General statement concerning taxonomy of inoceramid bivalves 53
Genus *Inoceramus* Sowerby, 1814 54
Inoceramus pictus lineage 54
Inoceramus pictus pictus Sowerby 55
Inoceramus pictus gracilistriatus Kauffman and Powell, n. subsp. 56
Inoceramus flavus pictoides Sornay 57
Inoceramus prefragilis prefragilis Stephenson 60
Inoceramus prefragilis stephensoni Kauffman and Powell, n. subsp. 62
Inoceramus ginterensis Pergament 64
Inoceramus tenuistriatus lineage 66
Inoceramus tenuistriatus? Nagao and Matsumoto 67
Inoceramus sp. aff. *I. tenuis* Mantell, s. l. 69
Inoceramus rutherfordi lineage 70
Inoceramus? *rutherfordi*? Warren, s. l. 70
Genus *Mytiloides* Brongniart, 1822 71
Mytiloides labiatus (Schlotheim) lineage 72
Mytiloides mytiloides (Mantell) 74
Mytiloides sp. transitional between *M. opalensis* (Böse) and *M. mytiloides* (Mantell) 78
Mytiloides opalensis (Böse) 79
Mytiloides sp. aff. *M. duplicostatus* (Anderson) 81
Mytiloides sp. cf. *M. submytiloides* (Seitz) 82
Mytiloides? *crippsi* lineage 84
Mytiloides? sp. aff. *M.*? *crippsi* (Mantell) 84
Family Entoliidae Korobkov, 1960 86
Genus *Entolium* Meek, 1865 86
Subgenus (*Entolium*): Emended 86
Entolium (Entolium) gregarium Kauffman and Powell, n. sp. 87
Genus *Syncyclonema* Meek, 1864 88
Syncyclonema? sp. 88
Family Gryphaeidae 88
Genus *Pycnodonte* Fischer de Waldheim, 1835 88
Subgenus *Phygraea*? Vyalov 88
Pycnodonte (*Phygraea*?) *newberryi* (Stanton) 89

Genus *Exogyra* Say, 1820 91
Exogyra sp. aff. *E. boveyensis* Bergquist 91
Family Ostreidae 92
Genus *Ostrea* Linné 92
Ostrea beloiti Logan 92
Ostrea? sp. aff. *O. anomioides* Meek 94
Class Scaphopoda Bronn, 1862 94
Family Dentaliidae Gray, 1834 94
Genus *Dentalium* Linné, 1758 94
Dentalium? sp. 94
Class Gastropoda 95
Family Acmaeidae Carpenter, 1857 95
Genus *Acmaea* Escholtz, 1833 95
Acmaea sp. aff. *A. cerillosensis* Johnson 95
Order Mesogastropoda 95
Mesogastropoda, sp. indet. 95
Family Pyramidellidae 96
Genus *Pseudomelania* Pictet and Campiche, 1862 96
Cf. *Pseudomelania* sp. 96
Family Siphonariidae 96
Genus *Anisomyon* Meek and Hayden, 1860 96
Anisomyon? sp. 96
Class Cephalopoda 97
Family Hamitidae Hyatt, 1900 97
Genus *Stomohamites* Breistroffer, 1940 97
Stomohamites simplex (d'Orbigny) 97
Stomohamites simplex cimarronensis Kauffman and Powell, n. subsp. 97
Genus *Hemiptychoceras* Spath, 1925 99
Hemiptychoceras sp. 99
Family Schloenbachiidae Parona and Bonarelli, 1897 99
Genus *Quitmaniceras* Powell, 1963 99
Quitmaniceras sp. cf. *Q. brandi* Powell, 1963a 99
Family Baculitidae Meek, 1876 100
Genus *Sciponoceras* Hyatt, 1894 100
Sciponoceras gracile (Shumard), 1860 100
Family Desmoceratidae Zittel, 1895 101
Genus *Desmoceras* Zittel, 1884 101
Desmoceras?, s. l., sp. 101
Subgenus *Pseudouhligella* Matsumoto, 1942 101
Pseudouhligella sp. 101
Family Acanthoceratidae Hyatt, 1900 101
Genus *Pseudocalycoceras* Thomel, 1969 101
Pseudocalycoceras spp. 1 and 2 101
Genus *Acanthoceras* Neumayr, 1875 102
Acanthoceras group "A" 103
Acanthoceras group "B" 103
Phylum Arthropoda 104
Class Cirrepedia 104
Acrothoracican barnacles 104
Notes on trace fossils 104
Phylum Protozoa 105
Class Rhizopoda 105
Family Heterohelicidae Cushman, 1927 105
Genus *Heterohelix* Ehrenberg, 1843 105
Heterohelix globulosa (Ehrenberg) 105
Family Schackoinidae Pokorny, 1958 106
Genus *Schackoina* Thalmann, 1932 106
Family Planomalinidae Bolli, Loeblich, and Tappan, 1957 106

Genus *Globigerinelloides* Cushman and Ten Dam, 1948 106
Globigerinelloides bentonensis (Morrow) 106
Globigerinelloides caseyi (Bolli, Loeblich, and Tappan) 107
Family Rotaliporidae Sigal, 1958 107
Genus *Hedbergella* Bronnimann and Brown, 1958 107
Hedbergella amabilis Loeblich and Tappan 107
Hedbergella delrioensis (Carsey) 108
Hedbergella planispira (Tappan) 108
Hedbergella portsdownensis (Williams-Mitchell) 109
Genus *Clavihedbergella* Banner and Blow, 1959 110
Clavihedbergella simplex (Morrow) 110
Genus *Praeglobotruncana* Bermudez, 1952 110
Praeglobotruncana stephani (Gandolfi) 110
Genus *Rotalipora* Brotzen, 1947 111
Rotalipora cushmani (Morrow) 111
Rotalipora greenhornensis (Morrow) 111
Genus *Anaticinella* Eicher, 1972 112
Anaticinella multiloculata (Morrow) 112
Family Marginotruncanidae Pessagno, 1967 113
Genus *Whiteinella* Pessagno, 1967 113
Whiteinella aprica (Loeblich and Tappan) 113
Genus *Marginotruncana* Hofker, 1956 114
Marginotruncana difformis (Gandolfi) 114
Plate section 115
Appendix 1. Description of measured sections . . . . . microfiche in pocket
References cited 141

FIGURES

Part 1

1. Map of westernmost part of Cimarron County, Oklahoma 8
2. Generalized stratigraphic and biostratigraphic chart for central and southern Western Interior 11
3. Stratigraphic columns showing Upper Cretaceous rocks in Cimarron County 13
4. Transgressive part of the Greenhorn Cyclothem showing relationship between stratigraphic sections in Cimarron County, Oklahoma, and north Texas 25
5. Biostratigraphic plot of all faunal elements studied 30

Part 2

1. Measured features on Cretaceous bivalves from northwestern Oklahoma 48
2. Interior shell morphology, members of the *Inoceramus pictus* lineage 58
3. Interior shell morphology, members of the *Inoceramus pictus* and *I. tenuistriatus* lineages . . 61
4. Stratigraphic distribution and principal morphologic features of taxa within the *Mytiloides labiatus* lineage 75
5. Diagrammatic representation of open coiled, slightly helical form of *Stomohamites simplex cimarronensis*, n. subsp. 96
6. Diagrammatic representation of partially preserved suture pattern of *Stomohamites simplex cimarronensis*, n. subsp. 98
7. Representation of a partially reconstructed specimen of *Hemiptychoceras* sp. 99

TABLES

Part 2
1. Summary of measurements for *Inoceramus flavus pictoides* Sornay 59
2. Summary of measurements for *Inoceramus prefragilis stephensoni*, n. subsp. 63
3. Summary of measurements for *Inoceramus ginterensis* Pergament 65
4. Summary of measurements for *Inoceramus tenuistriatus*? Nagao and Matsumoto 68
5. Summary of measurements for *Mytiloides mytiloides* (Mantell) 77
6. Summary of measurements for *Mytiloides opalensis* (Böse) 80
7. Summary of measurements for *Mytiloides* sp. aff. *M. submytiloides* (Seitz) 83
8. Summary of measurements for *Entolium (Entolium) gregarium*, n. sp. 86
9. Summary of measurements for *Pycnodonte newberryi* (Stanton) 90
10. Summary of measurements for *Ostrea beloiti* Logan 92
11. Summary of measurements for *Stomohamites simplex cimarronensis* n. subsp. 98

Acknowledgments

We express our gratitude to William A. Cobban, Norman F. Sohl (U.S. Geological Survey), Don L. Eicher (University of Colorado), and W. J. Kennedy (Oxford University, England) for partial or full review of the manuscript. L. B. Isham (Smithsonian Institution) and James Tolen (Indiana University) assisted in drafting, and Andrew Wynn and his staff (Smithsonian Institution) and George Ringer (Indiana Geological Survey) assisted with photography. Kauffman's field work was financed by a grant from the Smithsonian Research Foundation, and work on the Inoceramidae was financed by Grant GB-1627 from the National Science Foundation. Hattin's field work was financed by the State Geological Survey of Kansas, and Powell's work was financed in part by the University of Texas at Arlington. To these individuals and institutions we offer our sincere thanks.

Abstract

In the Oklahoma Panhandle, strata of Cenomanian-Early Turonian age contain invertebrate faunas, the study of which adds to the understanding of Western Interior biostratigraphy and furnishes important clues to the physical and chronologic relationships of Colorado and Kansas sections to those of north Texas. The section studied includes upper sandy beds of the Dakota Group; lower shale member, Thatcher Limestone Member, and upper shale member of the Graneros Shale; and the carbonate-rich Lincoln Member, Hartland Member, and Bridge Creek Limestone Member of the Greenhorn Formation. Stratigraphically and lithologically, the Graneros and Greenhorn of Cimarron County, Oklahoma, compare most closely with sections exposing the same units in westernmost Kansas; several bentonite and limestone beds are demonstrably common to the two areas. The Cimarron County section also compares closely with the Cenomanian-Lower Turonian section on the Model anticline of southeastern Colorado except that beds of limestone in the Bridge Creek Member are more brittle and less chalky in the latter area. Comparison of the Oklahoma section with that of north Texas confirms the general similarity of the two stratigraphic sequences but with marked diachrony of lithic equivalents in the lower part of the section due to the southeastward transgression during Cenomanian time. In Texas the Britton Member of the Eagle Ford Formation contains *Ostrea beloiti* Logan in association with a bentonite seam, which is believed correlative with the X bentonite of the Western Interior and which lies near the top of the Graneros Shale in western Oklahoma. Above this level, calcareous beds of the Britton Member reflect maximum transgression and equate chronologically to lower Greenhorn carbonates of Cimarron County.

Faunas of the Cimarron County section are a mixture of Western Interior and Gulf Coast taxa and include several cosmopolitan or intercontinentally distributed species. Uppermost strata of the Dakota Group yielded only possible worm and arthropod burrows. The lower shale member of the Graneros Shale is apparently barren of all but trace fossils. The Thatcher Limestone Member of the Graneros contains earliest representatives of *Inoceramus rutherfordi* Warren and is assigned a late Early to early Middle Cenomanian age. The lower part of the upper shale member of the Graneros is assigned to the upper part of the *I. rutherfordi* Range Zone, and the upper part is assigned to the *Ostrea beloiti* Range Zone, which is of Middle Cenomanian age. The upper shale member contains the first abundant foraminifera in the section, with benthonic forms predominating in the lower part and planktonic forms predominating in the upper part. Acanthoceratid ammonites used elsewhere to divide the upper Graneros into three widely traceable zones were not recorded in the Cimarron County section.

Biostratigraphic correlation and application of European stage terminology is readily accomplished in the Greenhorn Formation of Cimarron County. Correlation of the Lincoln Member is based on inoceramid bivalves that apparently represent the regional zone of *Dunveganoceras pondi* Haas. Nine species of planktonic foraminifera make their first appearance at the base of the Lincoln Member, but all are long-ranging forms. Ammonites, bivalves, and planktonic foraminifera are abundant in the Hartland Member and serve to define three local assemblage zones in the lower and middle parts of the member. The lowest zone is characterized by *Inoceramus flavus pictoides* Sornay and is assigned a Late Cenomanian age. The second and third Hartland assemblage zones are characterized by species of *Phelopteria;* associated bivalves and cephalopods suggest a Late Cenomanian age. Uppermost Hartland strata contain few recognizable macroinvertebrate fossils but contain a suite of typical Late Cenomanian foraminifera known from the Hartland of other areas.

The Bridge Creek Limestone Member of the Greenhorn Formation is divisible into three regionally widespread zones. The lowest of these is the well-known *Sciponoceras gracile* Zone, which contains a diverse assemblage of latest Cenomanian fossils and is recognized throughout the Western Interior region and northern Texas. In Oklahoma, the base of this zone marks the extinction level of a great majority of planktonic foraminifera that characterize the Lincoln and Hartland Members. The second assemblage zone marks the base of Turonian strata in Oklahoma and is characterized by *Mytiloides opalensis* (Böse), which corresponds to the zone of *Watinoceras coloradoense* Henderson in areas farther to the northwest. The uppermost part of the Oklahoma section is characterized by the zone of *Mytiloides mytiloides* (Mantell), which is equivalent to at least part of the *M. labiatus* zone of most American authors. The double-keeled foraminifer *Marginotruncana difformis* (Gandolfi) and *Neobulimina* sp. cf. *N. albertensis* (Stelck and Wall) first appear near the top of the *M. opalensis* Zone and thus coincide nearly with the base of the Turonian Stage as defined by the base of the *M. opalensis* Range Zone.

Study of fossil associations served to distinguish the following seven of the ecological assemblages that have been recognized elsewhere in the Western Interior by Kauffman:

1. An assemblage dominated by robust, byssate inoceramid and small ostreid bivalves is found in and adjacent to the Thatcher Limestone Member of the Graneros Shale and represents a high-energy, inner-shelf environment.
2. An assemblage dominated by large, flat, thin-shelled inoceramids is found in the upper part of the Graneros Shale and represents the quiet to gently agitated waters of the midshelf environment.
3. An assemblage dominated by robust, inequivalve, nonbyssate to slightly byssate inoceramids occurs in chalky shale, shaly chalk, calcarenite, and limestone beds of the Lincoln Member, Hartland Member, and Bridge Creek Limestone Member of the Greenhorn Formation and represents a diverse suite of offshore environments that range, depending on variations within this assemblage, from a quiet, foul-bottom situation to a gently agitated, normal-marine shelf.
4. An assemblage dominated by small, thin-shelled pectinids occurs in the lower and middle parts of the Hartland Member and represents relatively deep, quiet-water, offshore-shelf environments or mass-mortality concentrations on single bedding surfaces.
5. An assemblage dominated by subequivalve, slightly to moderately biconvex, nonbyssate to weakly byssate inoceramids occurs sparsely in the upper part of the Hartland and especially in the Bridge Creek Members and represents far offshore, deeper waters having gentle currents.
6. An offshore assemblage dominated by few ammonite taxa is found in the

upper part of the *Sciponoceras gracile* Zone, in the lower part of the Bridge Creek, and represents deeper, quiet-water, open-marine conditions.

7. An offshore assemblage dominated by diverse ammonite taxa is found in the Hartland Member and represents an open-marine, outer-shelf environment.

Dakota, Graneros, and Greenhorn strata described in Part 1 of this paper represent most of the transgressive part of the Greenhorn depositional cycle. Uppermost Dakota strata represent the transition from a marginal-marine to a marine environment along a fluviodeltaic coastline. Sandy shales in the lower member of the Graneros represent the inner sublittoral environment of moderate current and wave action at depths probably not exceeding 30 m (100 ft). The Thatcher Member of the Graneros represents a zone of offshore wave dissipation and depths of probably 15 m (50 ft) or less. The thin, even-bedded shales and the fauna of the upper member of the Graneros Shale represent deeper, quiet-water, farther offshore shelf environments. These were situated beyond the distributional range of coarse terrigenous detritus and experienced only occasional current scour on the sea floor, with water depths probably exceeding 45 m (150 ft). Even-bedded carbonate-dominated strata of the Lincoln and Hartland Members of the Greenhorn Formation represent deposition in quiet water, below wave base and well beyond the transport range of most terrigenous sand and silt. These strata and their associated pelagic and benthonic fossils reflect increased distance to the eastward-migrating shoreline and water depths probably exceeding 60 m (200 ft). Upward increase in amount of bioturbation in the Greenhorn section suggests progressively slower, probably interrupted sedimentation of fine carbonate oozes. The Bridge Creek Member represents the maximum stage of transgression that is expressed in the Cimarron County section; the strata imply deposition far from shore, beyond the distributional range of significant volumes of terrigenous detritus other than clay. In this member, pronounced alternation of purer, burrow-mottled limestones and little-burrowed, laminated chalk beds represent variations in rate of carbonate mud production or terrigenous detrital influx—probably the latter. While lower beds of the Bridge Creek were being deposited, optimum conditions prevailed for development of the pelagic biota, especially including ammonites, foraminifera, and coccolithophores. During deposition of upper Bridge Creek beds, the abundance of ammonites and benthonic macroinvertebrates declined markedly.

In Part 2 of this paper, macroinvertebrates and foraminifera of the Cimarron County section are documented, and most species are described in detail. Included in Part 2 are descriptions or discussion of 16 species of foraminifera, 10 species or species groups of ammonites, 5 species of gastropods, 9 species of noninoceramid bivalves, 13 species of inoceramid bivalves, and 1 species each of brachiopods, corals, and annelids. Trace fossils include acrothoracican cirripeds, *Crossopodia,* and several forms of burrow structures. New forms include the species *Phelopteria minuta* Kauffman and Powell and *Entolium gregarium* Kauffman and Powell. New subspecies described here include *Inoceramus prefragilis stephensoni* Kauffman and Powell, *Inoceramus pictus gracilistriatus* Kauffman and Powell, and *Stomohamites simplex cimarronensis* Kauffman and Powell.

PART 1

Stratigraphy and Paleoenvironments

Erle G. Kauffman
Donald E. Hattin
J. Dan Powell

INTRODUCTION

Purpose and Scope of the Study

The central part of the Western Interior Cretaceous marine basin—an area including southeastern Wyoming, central and eastern Colorado, western Kansas, northeastern New Mexico, and northwestern Oklahoma—contains one of the most complete and extensively studied records of large-scale cyclic sedimentation in the world. From numerous well-exposed sections in Colorado and Kansas, a detailed evolutionary history of epeiric marine environments and faunas has been constructed and analyzed. Three highly refined stratigraphic systems—physical, biostratigraphic, and geochronologic—are now available as tools for interpretation of geologic and biological events within the basin, as well as relationships with Cretaceous strata outside the Western Interior region.

Exacting lithostratigraphic studies such as those of Hattin (1962, 1965a, 1975) and Scott and Cobban (1964) are the basis for detailed interpretation of regional facies relationships (for example, see Kauffman and others, 1969), definition of major marine cycles, and construction of a basin-wide sedimentary model (Hattin, 1962, 1964; Kauffman, 1967a, 1970). Such studies constitute the physical stratigraphic framework for the basin.

At the same time, a highly refined biostratigraphic system has been developed, starting with comprehensive zonation of the Western Interior Cretaceous fauna by Cobban and Reeside (1952) and extending through lineage studies of ammonites (for example, see Cobban, 1951, 1962a, 1962b, 1963, 1964; Cobban and Scott, 1964, 1972; Reeside and Cobban, 1960), gastropods (Sohl, 1967), and bivalves (Kauffman, 1965, 1966a, 1967b, work in progress), as well as analyses of microfaunal assemblages from individual stratigraphic units (Eicher, 1960, 1965, 1966; Eicher and Worstell, 1970; Mello, 1969; Kent, 1968; Trexler, 1967). Kauffman and Kent (1968; Kauffman and Cobban, in prep.) organized a biostratigraphic revision of the Western Interior Cretaceous sequence integrating data from the entire biota, following the system outlined by Kauffman (1970). This revision permits temporal divisions as small as 0.08 to 0.32 m.y. per biostratigraphic zone. This level of temporal refinement is unparalleled elsewhere in the geologic column, yet still allows widespread correlation within and outside the basin. The present study indicates that this zonation can be extended readily into the Cretaceous sequence exposed in Cimarron County, Oklahoma.

The biostratigraphic system is the most practical tool for dating and correlation and is the backbone of a detailed system of chronostratigraphy for the Western Interior Cretaceous strata. In this paper, it is supplemented by two additional types of data—radiometric dates (Gill and Cobban, 1966; Kauffman, 1970; Obradovich and Cobban, 1973; Kauffman, 1977) and isochronous surfaces (for example, bentonites, certain concretion zones, and carbonate marker beds; Hattin, 1971). More than 400 bentonite levels and numerous other isochronous surfaces have now been defined in the Cretaceous section of Colorado and Kansas; many of these extend to northwestern Oklahoma. Most of these bentonites have been sampled for potassium-argon dating; at least 100 are datable. Thirty dates are now available for Albian through Lower Maastrichtian rocks, and many additional samples are in the process of analysis.

Most of this stratigraphic research has been carried out in Colorado, Kansas, Wyoming, Montana, and northern New Mexico—the middle of the Cretaceous basin where the most complete and best exposed marine record is preserved. Small, isolated, and (or) poorly exposed basinal outcrop areas, such as the Oklahoma Panhandle, and those of the basin margin—eastern Nebraska, western Iowa, Arizona, Utah, the eastern part of the Dakotas, and Idaho—have been largely neglected. However, the physical, biostratigraphic, and geochronologic systems elucidated for the central part of the Western Interior basin now make possible a broad spectrum of interpretive studies within this region and allow the detailed stratigraphy, depositional history, and regional facies relationships of these poorly known Cretaceous areas to be investigated. This is the first of these investigations.

During this recent period of evolution in stratigraphic sciences within the Western Interior, similar progress was made by investigators working the western Gulf Coastal Plain. Following careful documentation of the stratigraphy and of selected faunas (for example, Stephenson, 1941, 1952) and construction of an initial biostratigraphic framework (Stephenson and others, 1942), numerous workers (Young, 1963; Powell, 1965, 1971; Pessagno, 1969) attempted refinement of the biostratigraphy and evaluation of the regional physical and biological relationships within the Gulf Coast Cretaceous sequence. This work has made evident the close relationship between the Western Interior and Gulf Coast Cretaceous sections—in lithostratigraphy, sedimentary history, and biostratigraphy—to the extent that the Cretaceous strata of both areas clearly can be recognized as deposits of a single depositional basin (Kauffman, 1970, p. 614, 615, Fig. 1). In this regard, the general conclusions of Stephenson and Reeside (1938) are thus solidly reinforced. Differences between Gulf Coast and Western Interior rocks and faunas are primarily reflections of the chronology of cyclic sedimentation and the influence of warmer marine climates on the Gulf Coast.

For the past decade, we have been working independently on stratigraphic, paleontologic, and paleoenvironmental studies of Cretaceous rocks in Colorado (Kauffman), Kansas (Hattin), and Texas (Powell). We have been increasingly impressed with the considerable stratigraphic similarity between these areas. Prior to the present study, these relationships have never been summarized adequately. The advanced status of stratigraphic knowledge of the Cretaceous strata in these three areas now makes possible the detailed documentation of physical and biological relationships between the various parts of the Cretaceous System in the central Western Interior and the Gulf Coast, once the areas between them have been studied.

In seeking areas centrally situated between Colorado, Kansas, and Texas and presumably combining the stratigraphic characteristics of the three, two choices were obvious—eastern New Mexico and northwestern Oklahoma (Fig. 1). The eastern New Mexico sections have been the object of considerable study by Kauffman (in prep.), who found them closely similar to the Colorado-Kansas sequence. On the other hand, the marine Cretaceous rocks of northwestern Oklahoma were poorly known; the outcrop areas were small, discontinuous, and not noted for fossils. Yet this was the region most centrally situated to our three primary research areas, and it seemingly would combine their stratigraphic characteristics. We concluded that paleontologic and stratigraphic study of Cretaceous outcrops in Cimarron County, Oklahoma, would not only be a major contribution to our knowledge of poorly studied Cretaceous rocks of the Western Interior but also would furnish important keys to the physical and biostratigraphic relationships of the Colorado and Kansas sections to those of north Texas.

With these goals in mind, we met in Cimarron County, Oklahoma, in June 1968 and spent several days measuring, describing, and collecting fossils from the upper part of the Dakota Group, Graneros Shale, and Greenhorn Formation (Cenomanian through Early Turonian). This paper reports the results of our study. Part 1 includes description of the section, lithostratigraphy, zonation and correlation by macrofaunal and microfaunal criteria, regional correlation and facies analysis, and paleoenvironmental interpretations based on physical and biological data. Kauffman and Powell present the systematic paleontology upon which these interpretations are based in Part 2.

Previous Work

The first detailed stratigraphic work on Cimarron County, Oklahoma, was that of Rothrock (1925), who mapped and described a Cretaceous section consisting only of the Purgatoire Formation and overlying Dakota Sandstone (collectively the Dakota Group of this paper). He considered the Dakota Sandstone to be overlain throughout the area by rocks of Tertiary age. His section was incorporated without significant alteration in the index to Oklahoma stratigraphy published by Gould (1925). DeFord (1927) pointed out the need for revision of Rothrock's section, calling attention to the preservation locally of "Benton shales and limestones" between the Dakota and Tertiary strata. Acknowledgment of this addition to the known section was made by Gould (1929), who described the Benton as consisting of more than 18 m (60 ft) of "gray to drab limestone and shale, the limestone containing large numbers of the characteristic Benton fossil, *Inoceramus labiatus.*" He recorded the basinlike structure of the Benton rocks, gave details of the outcrop location, and noted that the Benton was surrounded on all sides by the Dakota. The northwestern part of Cimarron County was mapped again by Stovall (1938), who claimed erroneously (p. 587) that the Graneros Shale was the highest division of the Cretaceous section in the area, noting that the formation contains plesiosaur bones, shark teeth, and *Inoceramus* fragments. Later, Stoval (1942, p. 327; the names Greenhorn and Graneros were inadvertently transposed in this abstract) also established the presence of 9 m (30 ft) of "Greenhorn Limestone" (Bridge Creek Limestone Member of this paper), apparently applying the older concepts of Gilbert (1896) in the area; in this brief account the thickness of the Graneros was reported as 32 m (105 ft). The most recent geologic map of northwestern Cimarron County is that prepared by Schoff (1943), who depicted a large area of undifferentiated rocks of the Colorado Group (comprising the Graneros, Greenhorn, and Carlile formations of Western Interior usage) extending across parts of three townships, the largest part of this outcrop lying in T. 4 N., R. 2 E., C.M.,

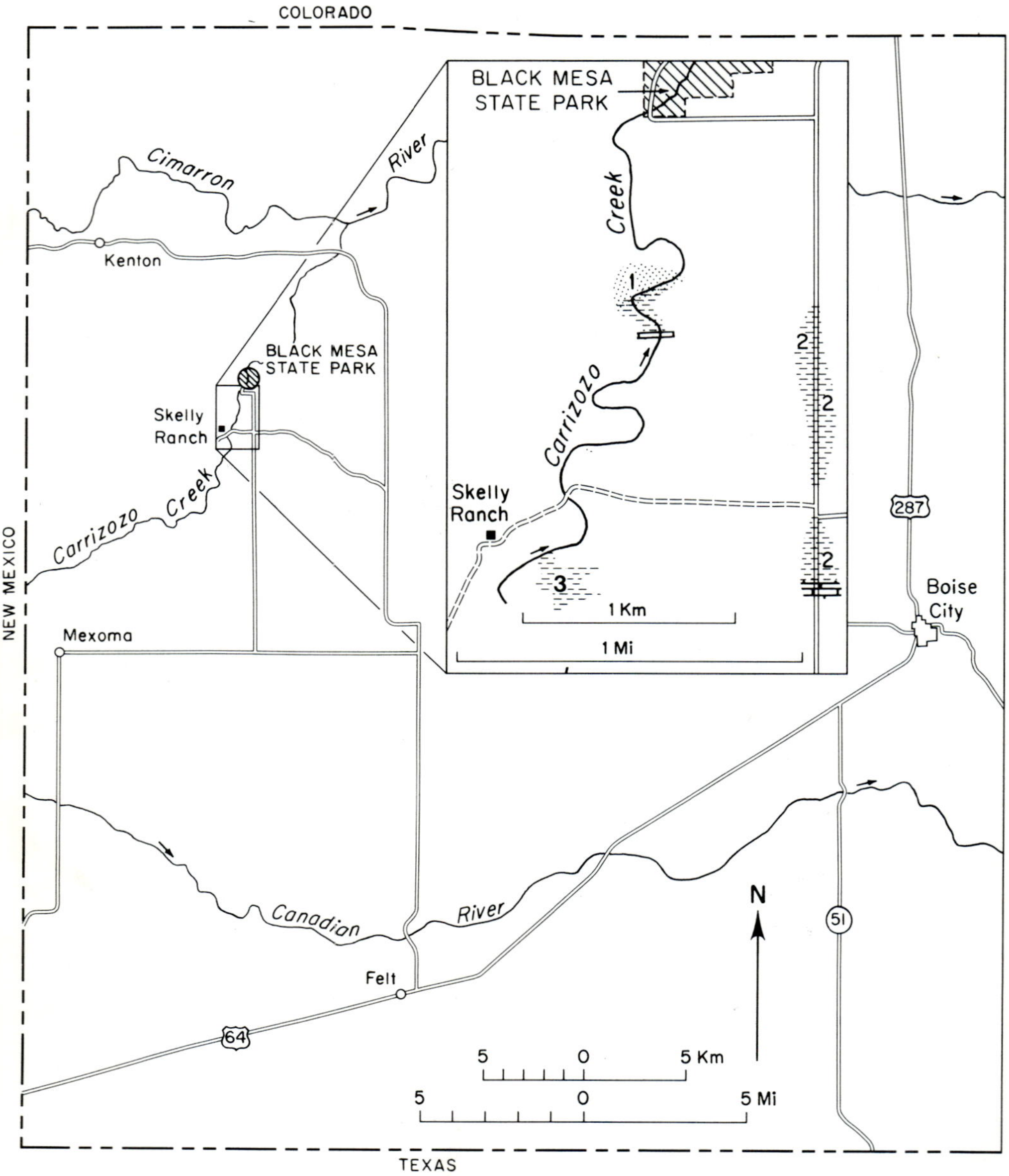

Figure 1. Map of westernmost part of Cimarron County, Oklahoma, showing (small rectangle) location of study area and (inset) location of measured sections. Locality of section 1—near center sec. 7, T. 4 N., R. 2 E. Locality of section 2—east line SE1/4 sec. 7, west line SW1/4 sec. 8, and west line NW1/4 sec. 17, T. 4 N., R. 2 E. Locality of section 3—NW1/4 sec. 18, T. 4 N., R. 2 E. Arrows show direction of stream flow; numbers indicate sections, with regional dip to south. Approximate area of outcrop indicated by following symbols: sandstone of Dakota Group shown by dots; clay and calcareous clay shales and shaly chalk (Graneros Shale through Hartland Member) shown by dashed lines; Thatcher (sec. 1) and Bridge Creek (sec. 2) Limestone Members shown by blocky pattern.

just east of the former town of Mineral. The map accompanied a bulletin in which Stovall (1943, p. 43–100) described the Mesozoic rocks in greater detail than any earlier author. He recorded (p. 96) thicknesses of approximately 20 and 21 m (65 and 70 ft), respectively, for the Graneros Shale and Greenhorn Formation, and he described a measured section extending 40.4 m (132.6 ft) above the top of the Dakota Group. This section contains several covered intervals, and Stovall's description lacks critical detail. Stovall (1943, p. 99) claimed that invertebrate fossils are poorly preserved in the two formations, but that great numbers of *Inoceramus labiatus* occur locally. He also mentioned *Ostrea congesta* and *Gryphaea?* sp. as being present at several localities and at different localities and horizons in the section.

In the years since Schoff's and Stovall's work was published, brief mention has been made, in many published works, of the Colorado Group rocks in Cimarron County. Since 1950, field excursions have been conducted in the area by at least two geological societies. Two of the guidebooks for these trips indicate a stop at a probable Greenhorn exposure on U.S. Highway 287, near the northeast end of the Cimarron County outcrop, but no stratigraphic details are given. We (Kauffman and others, 1969) presented a graphic section and generalized description of the Cimarron County section in our regional study of Cenomanian-Turonian facies across the Raton Basin. This was a preliminary survey of the detailed treatment presented here and, as far as we know, is the only significant work in this area since that of Stovall (1943).

RADIOMETRY AND ZONAL DURATIONS

The geochronology of the Western Interior Cretaceous sequence is perhaps the most refined pre-Pleistocene time scale in the world. More than 425 bentonite and (or) volcanic ash levels have been identified in Albian through Maastrichtian units of the North American Western Interior; a few sandstone units contain datable glauconite as well. Approximately 50 of these bentonites have now been dated, and many have been cross-checked between two samples or two minerals (sanidine and biotite). Of primary importance is the more or less even spacing of these dates through the section and their detailed correlation with biostratigraphic zones (see, for example, Gill and Cobban, 1966; Obradovich and Cobban, 1973, 1975; Kauffman, 1977). This allows calculation of a reasonably accurate *generalized time scale* for the Western Interior Cretaceous sequence (Kauffman, 1977) from which the rates and (or) duration of geologic and biological events (including biostratigraphic zones) can be accurately calculated; the spacing of actual dates from which this time scale is formed exercises good control on its general accuracy. Many of these dates can be fitted into the Cimarron County Cretaceous section on the basis of ash or bentonite stratigraphy itself or on the basis of the biostratigraphic relationships of the dated bentonite bed. The methodology by which this generalized time scale was developed and applied to subsequent zonal duration measurements in Oklahoma is briefly outlined below:

1. Detailed measurement and description of numerous stratigraphic sequences in the Western Interior determined the position, thickness, basic mineralogy, and lateral extent of strata containing datable minerals (usually bentonite or volcanic ash).
2. All potentially datable levels were properly sampled to provide material for analysis. Stratigraphically closely and evenly spaced samples were taken at more than 100 sampling levels where beds were greater than 2.5 cm (1 in.) in thickness

and where unweathered material could be obtained. Initially, material was obtained from a general area extending from central Colorado to southern Alberta. All major bentonite beds that could be traced widely were sampled at several distantly spaced localities to provide cross checks on the dating. Each sample was located precisely in a detailed physical stratigraphic system and within the general chronostratigraphic system of isochronous surfaces (ash falls, mass-mortality beds, concretion zones, and so forth) now being constructed for this region (Kauffman, in prep.).

3. Fossils were collected, wherever possible, directly above and below each sample used for radiometric dating until the assemblage was sufficiently large and diverse to determine the precise biostratigraphic position of the sample. Typically, fossil samples of biostratigraphically important groups were collected from sections where dated or datable bentonite, ash, or glauconite levels could be confidently identified so that the two sets of data could be closely related.

4. Dates were obtained from as many samples as possible. Where the material was available, analyses of different minerals (for example, biotite and sanidine in ashes or bentonites) were made and compared, and duplicate samples were run for a single stratigraphic level, preferably at different geographic localities. These formed the data base for construction of a generalized time scale, as described in steps 5, 6, and 7.

5. At levels which yielded more than one radiometric date, a single date was averaged from the mean values of all reliable dates obtained at that level. In some cases, it was beneficial to average two or more stratigraphically closely spaced samples from distinct levels. The resulting date was considered one of the key building blocks of the generalized time scale that was used for the measurement of evolutionary rates and zonal durations.

6. For a stratigraphic interval between two levels yielding radiometric data, equal time divisions between the bounding dates were calculated. These time divisions were usually calculated on the basis of events (species evolving within lineages, biostratigraphic zones, small-scale sedimentary cyclothems, and so forth) that *may* be nearly equal in duration over small spans of time and space; dates so calculated are pure assumptions, however, and should be used with care in measuring evolutionary rates or duration of zones. Usually it was the number of such events between known dates that determined the time divisions used in the generalized time scale. In the absence of a clear set of such divisions between known dates, the generalized time scale was usually constructed by dividing the sedimentary sequence (thickness) between dated levels equally into million-year increments. If two bounding dates were in fractions of millions of years (for example, 94.3 and 98.7 m.y. B.P.), the intervening stratigraphic sequence was proportioned into 0.1-m.y. divisions and then the points were identified that represented dates of whole million (or half million) years before present. In most Western Interior Cretaceous situations, the generalized time scale was filled out between dated levels by dividing the number of biostratigraphic units in each undated sequence into the difference between end dates of the interval and then from this, calculating the probable age of each biostratigraphic unit, or unit boundary. Such a scale is the best possible approximation of the age and duration of tangible geologic and biological units at this time.

7. Some ash levels (for example, the X bentonite) from which dates were obtained elsewhere also occur in the Cimarron County section. It was assumed that these bore the same real-time connotation; all bentonites encountered in Cimarron County were too much altered by weathering to yield reliable dates. Other dates were applied in Cimarron County on the basis of biostratigraphic zones associated with

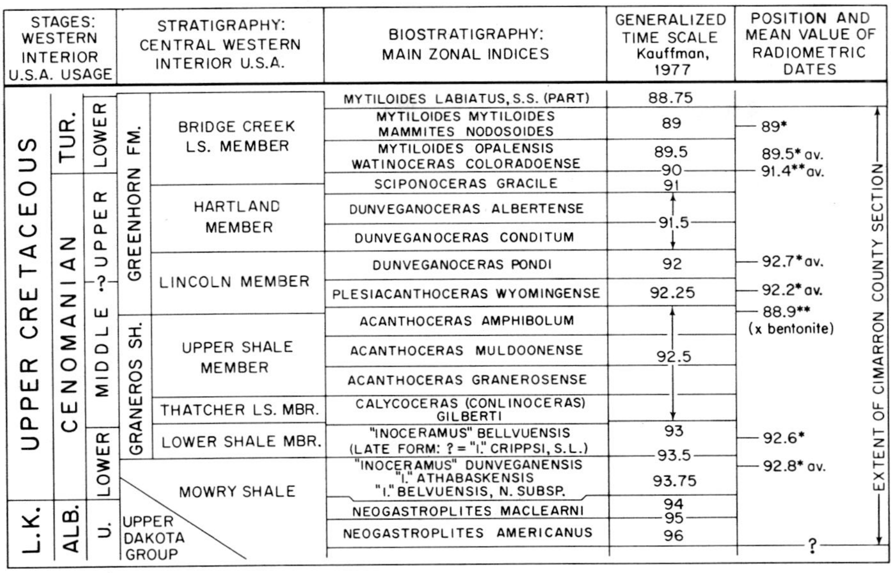

Figure 2. Generalized stratigraphic and biostratigraphic chart for central and southern Western Interior showing radiometric dates derived from Obradovich and Cobban (1973, 1975); Williams and Baadsgaard (1973, 1975); Folinsbee and others (1963), and Kauffman (1977). Key: ** = dates correlated on same bentonite and biostratigraphic zone; * = dates correlated on same biostratigraphic zone; av. = averaged from means of two or more dates.

dated materials from elsewhere in the Western Interior. Figure 2 shows the generalized time scale so correlated into Cimarron County and used there to measure zonal durations.

DESCRIPTIVE STRATIGRAPHY

In Cimarron County, Oklahoma, the exposed Cretaceous sequence includes (ascending) the Dakota Group (Aptian-basal Cenomanian(?)), the lower shale member, Thatcher Limestone Member, and upper shale member of the Graneros Shale (Cenomanian), and the Lincoln Member, Hartland Member, and Bridge Creek Limestone Member of the Greenhorn Formation (Cenomanian-Lower Turonian). The post-Dakota units were defined originally from localities in Colorado, Kansas, and Nebraska. Descriptions of the units, as characteristically developed in the type areas, are given by Hattin (1962, 1964, 1965a, 1965b, 1975) and Kauffman (1967a, 1969a). Waage (1953, 1955) described the stratigraphy of the Dakota Group in Colorado, where it is basically similar to the Dakota in northwestern Oklahoma. All of these units can be traced into Oklahoma through a series of intermediate exposures and with only minor changes of facies (Kauffman and others, 1969). The characteristics of these units in northwestern Oklahoma are summarized below. Their regional relationships and environmental implications are subsequently discussed and detailed descriptions of the measured sections are presented in Appendix 1 (see microfiche in pocket) and illustrated in Figure 3.

Dakota Group

During the present investigation, only the uppermost metre of the Dakota Group was examined in detail. Near the top of the group, cross-bedded sandstone, sandy

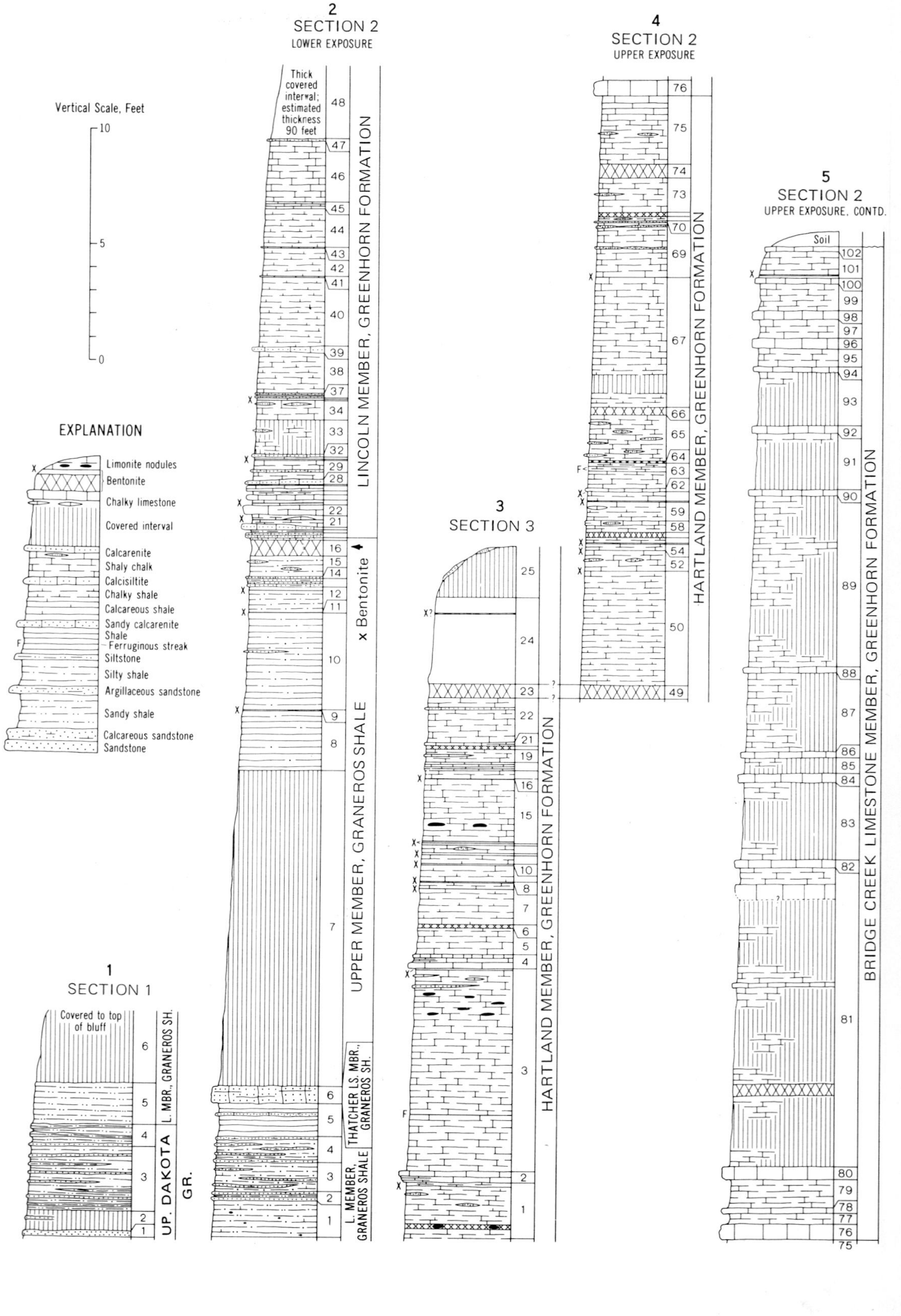
Vertical Scale, Feet
10
5
0
EXPLANATION
Limonite nodules
Bentonite
Chalky limestone
Covered interval
Calcarenite
Shaly chalk
Calcisiltite
Chalky shale
Calcareous shale
Sandy calcarenite
Shale
Ferruginous streak
Siltstone
Silty shale
Argillaceous sandstone
Sandy shale
Calcareous sandstone
Sandstone
1
SECTION 1
Covered to top of bluff
L. MBR., GRANEROS SH.
UP. DAKOTA GR.
2
SECTION 2
LOWER EXPOSURE
Thick covered interval; estimated thickness 90 feet
LINCOLN MEMBER, GREENHORN FORMATION
x Bentonite
UPPER MEMBER, GRANEROS SHALE
THATCHER LS. MBR., GRANEROS SH.
L. MEMBER, GRANEROS SHALE
3
SECTION 3
HARTLAND MEMBER, GREENHORN FORMATION
4
SECTION 2
UPPER EXPOSURE
HARTLAND MEMBER, GREENHORN FORMATION
5
SECTION 2
UPPER EXPOSURE, CONTD.
Soil
BRIDGE CREEK LIMESTONE MEMBER, GREENHORN FORMATION

shale, and lignitic beds are overlain by marine sandstone with oscillation ripple marks and well-preserved specimens of *Rhizocorallium*. Above the latter, the Dakota consists of about a metre of sandstone, siltstone, and shale representing the upward transition from marginal-marine beds to the open-marine beds of the Graneros Shale. In this transitional interval the sandstones are thin to very thin bedded, quartzose, fine to very fine grained, generally well cemented, and stained by limonite. The shales are medium to dark gray, very thin bedded, and contain much sand and silt. The uppermost part of the interval contains a little medium-gray, shaly siltstone. Shale and sandstone are interbedded through much of the interval, and all three rock types are intimately interbedded in the top 0.3 m (1 ft) of the Dakota. The Dakota-Graneros contact is placed on the basis of an abrupt upward decrease in sand content such that shale or mudstone, rather than sandstone, is the more conspicuous rock type.

Graneros Shale

Lower Shale Member. Exposed beds of the lower shale member consist predominantly of easily eroded, recessive shales and mudstones that generally are very sandy and (or) silty. Color of the rocks ranges from olive gray to medium dark gray where fresh and from dark grayish orange (10 YR 6/4) to dark yellowish orange (10 YR 6/6) where weathered. At the base, the rocks are noncalcareous, have limonite and minor jarosite stains along joints and bedding, and contain much selenite. In the upper part of the member, the rocks are weakly calcareous, and contain minor amounts of selenite; in one bed there are granule-sized limonite nodules. Some of the shales and mudstones contain minute specks or even large fragments of carbonaceous matter. Fecal castings and mostly horizontal burrow structures, as much as 3 mm in diameter, are common in the upper part of the member; rare fish scales and bones occur in one bed.

Although mudstones and shales predominate, lenses and thin to very thin, irregular to lensing beds of fine grained and very fine grained quartzose sandstone occur throughout the member. In two zones near the top, such sandstones are closely spaced and there form relatively resistant, ledge-forming beds each of which contains interbedded shale and mudstone. Some of these sandstones are thinly laminated to gently cross-laminated. The sandstones are weakly to strongly calcareous and are weathered light brown (5 YR 5/6) or, less commonly, very dark yellowish orange (10 YR 5/6). Sparse fish bone fragments, fecal castings, *Crossopodia*, and horizontal burrow structures were recorded from sandstone beds in the upper part of the member.

Poorly exposed shale beneath the Thatcher Limestone Member contains a single, very thin bed of dark yellowish-orange (10 YR 6/6) calcarenite containing fine quartz sand. This bed yields rare fish scales and mudstone-filled, oblique burrow structures.

On a hillside located 1.6 km (1 mi) south of the top of section 2 (location shown in Fig. 1), a Brunton compass traverse of the lower shale member suggested a total thickness of approximately 20 m (65 ft).

Thatcher Limestone Member. The Thatcher Limestone Member consists of 0.15 to 0.21 m (0.5 to 0.7 ft) of thin bedded to very thin bedded, well-cemented, slightly glauconitic, calcarenite containing fine quartz sand. This resistant, ledge-forming

Figure 3. Stratigraphic columns showing Upper Cretaceous rocks in Cimarron County, Oklahoma, based on our detailed measured sections (described in App. 1). Locations are given in Figure 1. Numbers are unit numbers used in text.

member is highly jointed and weathers into irregular slabs that lie scattered on slopes below the exposure. In places a prominent topographic bench is developed on the Thatcher. The rock is weathered dark yellowish brown (10 YR 4/2) to yellowish brown (10 YR 5/2), has been calichified along joints, and contains scattered secondary nodules of alunite. The principal fossils are calcitic *Inoceramus* prisms, of which the rock is predominatly composed. Also present are valves or valve fragments of *Ostrea* and *Inoceramus*, fish scales and bones and bone fragments, and phosphatic coprolites. Petrographically, the rock is a partially recrystallized, *Inoceramus*-rich biosparite.

Upper Shale Member. The upper part of the Graneros consists largely of soft, slightly to very silty clay shale that is noncalcareous except in the uppermost 0.3 to 0.6 m (1 to 2 ft). Little-weathered rock is medium to dark gray or light olive gray (5 Y 6/1) to olive gray (5 Y 4/1). In section 2, most of the exposed rock is weathered yellowish brown (10 YR 5/2) to moderate brown or dark grayish orange (10 YR 6/4). Some shale units are stained dark yellowish orange by limonite. Most of the shale is fissile; some is laminated. Part of the shale contains fine specks of carbonaceous matter.

The top half of the upper shale member contains two thin to very thin beds of very fine grained silty, and in places clayey, noncalcareous sandstone. In the top 0.3 m, very thin lenses of calcarenitic quartzose sandstone occur. Most of the sandstone is weathered dark yellowish orange and is very thinly laminated to gently cross-laminated. The sandstone contains scattered remains of oysters, *Inoceramus*, and a few fish bones, scales, and teeth.

This member also contains four beds of nearly white bentonite, all of which are at least partially stained dark yellowish orange by limonite. Only one of these beds, the 21-cm- (0.7-ft) thick X bentonite, is more than 1.2 cm (0.04 ft) thick, and one highly ferruginous bentonite seam is discontinuous. Granular selenite occurs along the margins or within two of the seams, and some of the bentonite contains a small amount of noncalcareous siltstone.

Near and at the top of the member are two very thin beds of calcarenite and calcisiltite, respectively. The lower is a lensing bed composed largely of disarticulated oysters and some quartz sand. The calcisiltite is soft, recessive, and contains abundant *Inoceramus* prisms and fragments. These thin carbonate beds and accompanying calcareous shale reflect the transition from the predominantly terrigenous deposits in the Graneros Shale to the carbonate-rich deposits in the overlying Greenhorn Formation. In section 2, the upper shale member is 7.22 m (23.7 ft) thick.

Greenhorn Formation

Lincoln Member. Exposed beds of the Lincoln Member of the Greenhorn Formation consist largely of weathered, soft, recessive shaly chalk in layers ranging from 0.03 to 0.94 m (0.1 to 3.1 ft) in thickness. Coloration is mostly grayish orange to dark grayish orange (10 YR 6/4) or dark yellowish orange owing to oxidation of iron compounds. Planktonic foraminifera and *Inoceramus* prisms are concentrated locally, and scattered specimens are probably responsible for the sandiness of most beds. Nearly white fecal (?) pellets are evident in some layers. Macroinvertebrate remains include scattered fragments of *Inoceramus*.

Calcarenite is the characteristic, if not volumetrically preponderant, rock type in the exposed part of the member, and occurs as numerous thin to very thin, commonly lensing beds distributed irregularly through the Lincoln. Most of these are well-cemented, ledge-forming beds that litter the underlying slopes with platy slabs where eroded. The partly weathered rock is pale yellowish brown to dark yellowish brown; most, however, is weathered very pale yellowish brown (10 YR 6/4), grayish orange, or dark yellowish orange with limonite stains. These calcarenites commonly are thinly laminated to gently cross-laminated. Some layers are composed predominantly of planktonic foraminiferal tests; others are composed mainly of *Inoceramus* prisms. Macroinvertebrate fossils include valves and fragments of *Inoceramus* and oysters, sparse ammonite molds, rare burrow structures, and sparse fish bones, scales, and teeth. Samples of these rocks were determined in thin section to be foraminiferal and (or) *Inoceramus*-prism biosparites.

Of lesser importance is chalky limestone that occurs as several thin to very thin beds from 0.9 to 11.3 cm (0.03 to 0.37 ft) thick and distributed irregularly through the exposed part of the Lincoln. Although most of these beds are well indurated and form ledges, one bed is soft and crumbly where weathered. Two of the beds are extensively jointed. White ellipsoidal pellets were observed in some of the beds; all of the rocks are micrograined. The least weathered rocks are medium gray, but nearly all are highly weathered to grayish orange. Some are much stained by limonite. These rocks contain few fossils, but scattered *Inoceramus* fragments and fish scales, bones, and teeth were recorded in some of them. In thin section, samples of this rock type were determined to be foraminiferal microsparites or, rarely, biomicrosparite.

Soft, unusually silty, nearly white bentonite occurs in the Lincoln Member in three seams ranging from 0.9 to 3.0 cm (0.03 to 0.1 ft) in thickness. All are dark yellowish orange where stained by limonite.

Uncommon Lincoln rocks include grayish-orange (after weathering), well-laminated, soft, recessive calcareous shale containing silt-sized quartz grains that occurs just above the basal calcarenite. There is also one thin layer of soft, dark yellowish-orange, foraminiferal chalk 0.6 m (2.1 ft) above the base of the member.

In section 2, the exposed part of the Lincoln Member is 5.27 m (17.3 ft) thick.

Hartland Member. Strata of the Hartland Member of the Greenhorn Formation consist mostly of soft, recessive, thin to very thinly laminated shaly chalk that forms beds from 3 cm (0.1 ft) to as much as 1.5 m (5 ft) in thickness. These rocks are gritty owing to the abundance of calcareous silt and very fine sand, probably consisting largely of foraminifera and *Inoceramus* prisms; they are speckled essentially throughout by white, sand-sized ellipsoidal fecal (?) pellets. The nonweathered rock (only a small amount of the total) is olive gray (5 Y 4/1) or medium gray; partially weathered rock is mostly light olive gray; more extensively weathered rock is predominantly pale grayish orange (10 YR 8/4), grayish orange, dark grayish orange (10 YR 6/4), or dark yellowish orange. Limonite staining is common. Powdery gypsum occurs along some bedding planes in the lower part of the exposed section, as well as a little granular selenite. In some stratigraphic intervals, highly chalky rock alternates with thin units of chalky shale, but the shale facies is of minor importance volumetrically. The most common fossils are *Inoceramus* valves, mostly fragmentary, and fish bones and scales. Shaly chalks have also produced a few ammonite molds and noninoceramid bivalves. Tests of planktonic foraminifers are conspicuous in several units.

The Hartland contains, mostly in its lower half, a few moderately hard, commonly brittle, ledge-forming chalky limestones in beds from 1 to 15 cm (0.04 to 0.5 ft) thick. These beds are thin to very thin bedded, and some are also thinly laminated. Some of the limestone is speckled by small ellipsoidal fecal (?) pellets. Sparse, relatively nonweathered rock is medium dark gray, but most is weathered grayish to pale grayish orange (10 YR 8/4), pale yellowish brown, light olive gray, or yellowish gray. Limonite partially stains most beds dark yellowish orange. The most common fossils are sparse whole specimens and scattered fragments of

Inoceramus and fish scales. Sparse to moderately abundant, noninoceramid bivalves, locally abundant planktonic foraminifers, and fish bones are also present. In thin section, samples of these rocks were determined to be mostly poorly laminated foraminifera- and (or) pellet-rich microsparites.

Next in volumetric importance is calcarenite that occurs predominantly as very thin lenses in shaly chalk mostly in the lower part of the exposed section. The lenses range up to 0.9 cm (0.03 ft) thick; rare lensing beds of calcarenite are as much as 6 cm (0.2 ft) thick. Although the uppermost recorded calcarenite has a chalky matrix and is not very hard, the rest of these rocks are well cemented by sparry calcite; they are hard, brittle, and resistant to erosion and form minor ledges on the exposures. When freshly broken, the calcarenites commonly emit a petroliferous odor. Some calcarenites are composed dominantly of foraminifera; others consist dominantly of *Inoceramus* prisms. A majority of the calcarenite beds and lenses contain fragments of *Inoceramus*. A few fish bones, burrow structures, and ammonite molds constitute the other fossils recorded from this rock type. Petrographically, these rocks are *Inoceramus*-prism and (or) foraminifer-rich biosparites.

Soft, nonlaminated chalk is very minor in the Hartland and occurs as very thin foraminifer-rich lenses that are weathered grayish orange to very pale orange.

Bentonite seams are numerous in the Hartland and are scattered throughout the section in layers as little as 0.3 cm (0.01 ft) thick to as much as 18 cm (0.6 ft) thick. The fresh rock is nearly white, but most seams are stained dark yellowish orange, or far less commonly light brown (5 YR 5/6), by iron oxide. About half the bentonites contain noncalcareous silt. Seams in the lower part of the exposed section are commonly accompanied by granular gypsum that borders, divides, or is scattered through the seams and makes up most of the thickness of some beds. A few seams presumed to have been bentonitic are now composed wholly of iron oxide.

The Hartland Member contains more bentonite and less calcarenite than the Lincoln Member. The Lincoln has more and thicker beds of calcarenite than the Hartland.

In sections 1 and 2, the exposed parts of the Hartland Member are a total of 15 m (49.6 ft) thick.

Bridge Creek Limestone Member. Like the underlying part of the Greenhorn Formation, the Bridge Creek Limestone Member consists mainly of soft, recessive shaly chalk in beds from 0.15 m (0.5 ft) to as much as 2.3 m (7.5 ft) in thickness. Nearly all of this rock is thinly laminated, some is speckled by white, ellipsoidal fecal (?) pellets, and most is made gritty by calcareous silt that probably consists largely of microfossils. The chalk is weathered and poorly exposed, but the least weathered rock is light olive gray to very light gray. Typically the chalk is much weathered, the predominant colors being grayish orange and pale grayish orange (10 YR 8/4) with dark yellowish-orange limonite stains in most layers. Because of poor exposure and extensive weathering, these rocks yielded only a few macrofossils including sparse inoceramid fragments and rare fish scales. Scattered to locally abundant tests of foraminifera occur in most beds.

The characteristic rock type of this member is micrograined limestone and chalky limestone that occurs in massive, usually burrow-mottled single-bed units that are 6.1 cm (0.2 ft) to as much as 21 cm (0.7 ft) thick. All of these rocks are hard, brittle, resistant, and form at least minor ledges. Most of the beds are well jointed and break into small or large rectangular blocks. A few beds have an undulatory upper surface. Two beds contain minute seams of secondary calcite, and a few beds of both chalky and nonchalky limestone are semicrystalline. The nonchalky limestones are concentrated in the upper part of the member, in the interval equivalent to the Jetmore Member of the Greenhorn Formation in Kansas (approximately equivalent to the lower Bridge Creek Member of this paper). A few of the chalky limestone beds are little weathered and in part preserve medium light to medium dark gray coloration. More typically, both kinds of limestone are weathered yellowish gray, very pale orange, pale grayish orange (10 YR 8/4), or grayish orange. Most of the nonchalky limestone beds are stained dark yellowish orange by limonite. In addition to the general burrow mottling of these rocks, slender limonite- and (or) calcite-filled cylindrical burrows are preserved in several beds. Whole and broken specimens of inoceramids are sparse to abundant in various beds, with greatest concentrations in the three youngest beds. Four beds of chalky limestone in the lower part of the member contain relatively common molds of ammonites, but these fossils are rare in the nonchalky limestones. Foraminifers are conspicuous in several beds but are abundant only in one limestone unit (sec. 2, unit 86). In thin section, most of these rocks were determined to be foraminiferal micrites and microsparites or biomicrites and biomicrosparites. *Inoceramus* prisms and fragments were found in most thin sections studied, but in only one sample did these fossils exceed the foraminifer tests in abundance. The microsparites and biomicrosparites are the neomorphosed equivalents of micrite and biomicrite, respectively. Several thin sections contained sparry calcite-filled fractures, and one thin section contained silicified fragments of *Inoceramus.*

A major bed of bentonite, 15 cm (0.5 ft) thick and containing noncalcareous silt, lies approximately 1.8 m (6 ft) above the base of the member. Where fresh, this bed is medium dark gray but is mostly stained dark yellowish orange by iron oxide.

The Bridge Creek Member is characterized by the abundance and close spacing of resistant thin to medium beds of limestone and chalky limestone. Unlike the underlying members, the unit has no calcarenite. In section 2, the exposed part of the Bridge Creek is 13 m (42.8 ft) thick.

REGIONAL STRATIGRAPHIC RELATIONSHIPS

Comparison with Kansas Section

Hattin (1965a, 1975) has completed detailed stratigraphic studies of the Graneros Shale and Greenhorn Formation in Kansas. The rock types of the Graneros Shale and the members of the Greenhorn Formation are comparable in gross aspect to those of equivalent units in Cimarron County, Oklahoma, as well as in specifically correlative small stratigraphic intervals.

In both areas, the predominant rock type of the Graneros is gray, noncalcareous, silty and (or) sandy clay shale that weathers into flakes. The lower part is characterized by numerous thin to very thin beds of fine-grained sandstone. Higher in the Kansas section, sandstones are progressively less common; in their place are thin to very thin lenses and beds of calcarenite that commonly contain quartz sand or silt. Shale near the top of the formation is weakly calcareous in several Kansas localities, and contains sparse terrigenous sand and silt. The Graneros and Greenhorn are lithologically transitional at two or three localities. The Kansas section contains little mudstone of the kind that lies near the top of the lower shale member of the Graneros in Cimarron County. In central Kansas the Graneros Shale is much thinner (average thickness for 20 sections, 9.3 m [30.5 ft]) than in Cimarron County and contains no demonstrable equivalent of the Thatcher

Limestone Member. However, a prominent, tightly cemented calcarenite lying approximately 12 m (40 ft) below the top and 6.1 to 6.7 m (20 to 22 ft) above the base of the Graneros in Kearny County, Kansas (Bass, 1926, p. 72), probably represents the easternmost exposure of the Thatcher. A thick bentonite marker bed (X bentonite of many authors) lies near the top of the Graneros at most Kansas exposures (Hattin, 1965a, 1968). This bed is the same as unit 16 of section 2 and, as in Kansas and southern Colorado, is associated with thin beds of calcarenite and biostromal limestone containing an abundance of *Ostrea beloiti* Logan.

In Kansas, the Lincoln and Hartland Members of the Greenhorn Formation are well defined. The Lincoln Member consists mainly of dark olive-gray shaly chalk through which are scattered numerous thin to very thin beds and lenses of skeletal calcarenite and seams of bentonite. The member also contains scattered thin to very thin beds of relatively soft, olive-gray, chalky limestone that weathers most commonly to grayish orange. In Cimarron County the Lincoln section is lithologically comparable to that in Kansas, but the chalky limestone beds are much harder. In Kearny County, Kansas, the Lincoln is 4.8 m (16 ft) thick compared to at least 5.2 m (17.3 ft) exposed in Cimarron County where additional Lincoln beds may be included in the thick, overlying covered interval that forms unit 48 of section 2. The upward decrease in number and thickness of limestone beds, as well as the absence of evidence indicating concentration of calcarenites or chalky limestones in the upper Lincoln of Cimarron County, suggests that the Lincoln may be transitional with the overlying Hartland as in Washington and Ford Counties, Kansas.

In Kansas the Hartland Member consists predominantly of dark olive-gray shaly chalk containing a variable number of thin, burrow-mottled chalk and chalky limestone beds and several thin to very thin beds of bentonite, two of which reach thicknesses as great as 18 to 21 cm (0.6 to 0.7 ft). Calcarenite is uncommon in the Kansas Hartland, but a few very thin lenses occur locally in the unit. At the type locality in Kearny County, the Hartland is 13.5 m (44 ft) thick. The Cimarron County section is closely comparable lithologically but contains a greater abundance of calcarenite lenses than most sections of the Kansas Hartland. At least one marker bed can apparently be traced between the two areas. Unit 4 in section 3 is lithologically and faunally like a similar limestone bed in the type Hartland of Kearny County, Kansas. In both places the rock is a coarse-grained microsparite that weathers to small plates, has abundant pellets, and is underlain by a very thin seam of bentonite. Hartland bentonites are numerous above this bed in both areas. In Kearny County the bed lies 7.6 m (25 ft) below the top of the member; in Cimarron County this bed lies 11.3 m (37 ft) below the top.

The Lincoln-Hartland interval in Cimarron County, including a 27.5-m (90-ft) covered interval (unit 48, sec. 2), is approximately 40 m (133 ft) thick. This is more than twice the thickness present in Kearny County, Kansas (170 km [105 mi] to the northeast), nearly 2.5 times the thickness in nearby Baca County, Colorado (McLaughlin, 1954, p. 115, 116), and nearly 3 times the thickness reported by Bass and others (1947) for the occurrence in the Model anticline that lies approximately 135 km (85 mi) to the northwest of locality 2. This is the only major thickness discrepancy between Cimarron County and adjacent sections. Possible explanations for this discrepancy are (1) the Cimarron County area was more actively subsiding and received more sediment than areas to the west, northwest, north, and northeast or (2) owing to structural complications in unit 48 of section 2 (the covered interval), the calculated thickness of this unit measured along 1 km (0.6 mi) of roadway, is substantially in error. The second alternative seems more plausible because our measured thickness for the underlying Graneros Shale

(24.1 to 27.5 m [79 to 90 ft]) is compatible with the westward-thickening trend that exists between Kearny County, Kansas (18.3 to 19 m [60 to 62 ft]), Baca County, Colorado (26 m [85 ft]; McLaughlin, 1954), and the Model anticline of Las Animas County, Colorado (28.2 m [92 ft]). Furthermore, the exposed part of the overlying Bridge Creek Limestone Member in Cimarron County is almost exactly the same thickness as that for the equivalent interval in Hamilton County, Kansas (13.5 m [44 ft]).

The Bridge Creek Limestone Member of the Cimarron County section contains numerous beds of burrow-mottled limestone and chalky limestone. Similar beds are characteristic of the type Bridge Creek in Hamilton County, Kansas, and many of these beds can be traced for hundreds of kilometres in the area between Washington County, in northern Kansas, to Pueblo County, Colorado (Hattin, 1971). Three of these widespread markers can be identified confidently in section 2 (Fig. 3), and these serve as a basis for comparing the remaining parts of the Bridge Creek with that in the type section, lying less than 160 km (100 mi) to the northeast. The Cimarron County section contains fewer limestone beds, but those present are closely similar in texture, structure, composition, and thickness to those of the type section. However, the limestone beds in the upper half of the Oklahoma Bridge Creek are less chalky and more brittle, possibly owing to weathering. Unit 76 of section 2, lying at the base of the member, is correlated with the basal bed of the Bridge Creek Member at the type locality (Hartland bed 1 of Hattin, 1971). The lower chalky limestone bed in unit 81 contains a fossil assemblage that is typical also of a widespread, thin, chalky limestone bed in the lower part of the type Bridge Creek (Hartland bed 2 of Hattin, 1971). As in Kansas, the limestone bed lies shortly above a prominent seam of bentonite; this fact, coupled with the fossil evidence, suggests that the limestone is the same as Hartland bed 2 (Hattin, 1971). Unit 100 of section 2 is correlated with a chalky limestone bed lying above the middle of the type Bridge Creek (Jetmore bed 10 of Hattin, 1971). In both localities this limestone bed contains numerous noncrushed specimens of *Mytiloides mytiloides* (Mantell) (= *Inoceramus labiatus* of most American authors), is harder than the subjacent chalky limestone beds, and is associated with a thin overlying seam of bentonite. In both areas, olive-gray shaly chalk beds separate the limestone units of the Bridge Creek. The measured section of the Bridge Creek Member in Cimarron County, including units 76 through 102, is approximately 13.2 m (43 ft). The stratigraphically equivalent section in the type Bridge Creek of Hamilton County, Kansas, is 13.5 m (44 ft). The type Bridge Creek includes, above the equivalent of our unit 102, 9.8 m (32 ft) of strata that have been eroded from the Cimarron County section.

In summary, the exposed parts of the Cimarron County section compare well with the section in Kearny and Hamilton Counties, Kansas; the thicknesses of comparable parts of the Cimarron County section are equal to or as much as 50 percent greater. The poorly exposed part of the Bridge Creek contains an apparently smaller number of limestone beds than is normal in areas to the northwest and northeast, and the thickness of the covered interval separating the exposed parts of the Lincoln and Hartland at locality 2 may be considerably in error owing to structural subtleties.

Comparison with Southern Colorado Sections

Stratigraphic units closely comparable to those in the Upper Cretaceous section of Cimarron County are well exposed along most of the southern frontal ranges (Wet Mountains, southern Sangre de Cristo Range) in Colorado and around low arches and anticlines in southeastern Colorado. In southeastern Colorado, especially around the Model anticline in Las Animas County (138 km [85 mi] northwest of the Cimarron County outcrop), the section is nearly identical to that in the Oklahoma Panhandle. Westward, in Huerfano County, Colorado, thicknesses change, less calcareous facies develop in parts of the section, the Thatcher Limestone Member pinches out or passes laterally into one or two concretion zones, and a thin tongue of the Mowry Shale wedges between the Graneros Shale and uppermost Dakota Group.

Comparison with the Model Anticline. Although the Graneros Shale is only partially exposed in the Oklahoma Panhandle and in Las Animas County, Colorado, the formation is very similar in both areas. At the top of the Dakota, hard ferruginous sandstones and interbedded sandy shales grade upward into a series of dark gray to dark olive-gray silty and sandy shales containing lesser amounts of shaly sandstone and thin siltstone beds that characterize the lower part of the Graneros Shale. In the lower Graneros, sandstone and siltstone beds are rarely more than 8 cm thick and are easily weathered; this is in sharp contrast to the hard, ledge-forming Dakota units below. In both areas this rock type persists to the top of the lower shale member, and thin brown sandstone beds, many of them burrowed, occur within a few feet of the Thatcher Member. The number of sandstones and the amount of sand in the shale decrease upward in both area. In Las Animas County the middle and upper parts of the lower shale member of the Graneros Shale contain considerable thicknesses of mudstone, clay shale, and only slightly silty clay shale; similar rocks dominate the upper part of the lower shale member in the Cimarron County section.

According to Bass and others (1947) the lower shale member is 11 m (35 ft) thick on the Model anticline; in the same area Kauffman and Hattin determined the thickness to be 19 m (62 ft; previously unpub. data). This is only slightly less than the maximum thickness (range, 16.9 to 21.3 m [55 to 70 ft]) that we measured in Cimarron County in areas of considerable soil cover but where no structural complications were suspected and dips were apparently uniform. In both areas the total thickness of the Graneros is comparable, being 25.4 m (87 ft) in Cimarron County and, depending on interpretation of the Graneros-Greenhorn contact, 29 to 46 m (96 to 105 ft) in Las Animas County. This thickening trend continues westward to the Front Range in Colorado where the Graneros is 40 m (130 ft) thick along the southern part of the Front Range (Kauffman, 1967a, p. 75). Slight east-west variations in thickness of the two shale members of the Graneros may relate to this general trend, to lateral facies change, to local uplifts and depressions in the basin, and (or) to shifts in the relative stratigraphic position of the lenticular and concretionary Thatcher Limestone Member that separates them (Kauffman and others, 1969, Fig. 4).

The Thatcher Limestone Member of the Graneros Shale extends from Cimarron County to the Model anticline with little change. At the type section, Bass and others (1947) reported about 2 m (7 ft) of Thatcher, including a lower 10- to 15-cm-thick (4- to 6-in.) limestone bed, 1.5-m-thick (5-ft) shale interval and an upper 46-cm-thick (18-in.) limestone bed. Re-examination of this section by Hattin and Kauffman in 1968 revealed that the Thatcher consists only of a single calcarenite bed of variable thickness that appears to consist of two beds at the type section because of vertical displacement along a small fault. In other areas a discontinuous quartzose sandstone bed in the lower Graneros, as much as 0.5 ft thick, lies 6.5 ft below the Thatcher, but we recorded nowhere more than one bed having Thatcher lithology. Cobban and Scott (1972, p. 10) agreed with this interpretation and redefined the type Thatcher so as to include only a single bed of hard, brittle limestone. Two zones of limestone concretions containing fossils approximately of the same age as those in the Thatcher occur in Huerfano Park, Colorado, and were correlated tentatively with this member by Kauffman and others (1969).

The Thatcher Limestone Member is thicker (15 to 46 cm [0.5 to 1.5 ft]) and finer grained in the type area than it is in Cimarron County (11 to 15 cm [0.35 to 0.5 ft]). Bedding is somewhat irregular and thickness is variable in both areas. Although a single bed persists locally, it can be demonstrated through parts of southeastern Colorado that the limestone beds of the Thatcher are regionally lenticular and are represented locally as concretionary lenticular masses. *Inoceramus* and oyster-shell debris, fish scales, and bone fragments are the principal organic components.

The upper shale member of the Graneros thickens gradually from Cimarron County northwestward to the Model anticline but is lithologically similar in both areas, being composed dominantly of medium to dark gray and olive-gray clay shale and mudstone interbedded with slightly to moderately silty clay shale. Thin brown sandstone and siltstone beds are developed sparingly at some localities.

In Cimarron County the Graneros-Greenhorn contact is fairly typical of this contact throughout much of the central Western Interior. The upper metre of the Graneros is more or less calcareous and contains thin beds of calcisiltite and scattered very thin beds of calcarenite. A prominent bentonite marker bed, the X bentonite, lies near the top of the member, and lies at or near the contact throughout the central part of the Western Interior. Throughout the region the base of the overlying Lincoln Member is drawn at the base of the first prominent ledge-forming calcarenite bed or group of such beds. This usually lies just above the X bentonite but in some cases lies below it within a distance of 1 m (3 ft). Along the northern Front Range and in the Denver Basin area of Colorado, Cobban (1974, written commun.) suggested that the base of the Lincoln be drawn at the base of the X bentonite inasmuch as the shales become strongly calcareous at approximately this level, and the first calcarenites lie 1 to 2 m above the bentonite in some sections.

In their study of the Model anticline, Bass and others (1947) defined the Graneros-Lincoln contact as lying 4 m (13 ft) above the top of the X bentonite. They noted noncalcareous gray clay shale in this 4-m-thick interval and designated as the base of the Lincoln a 1.5-m-thick (5-ft) series of ledge-forming calcarenite beds. They noted also a 0.3-m-thick (1-ft) limestone and oyster-coquina bed directly overlying the X bentonite. The latter is a widespread *Ostrea beloiti* biostrome that has been recorded at the base of the Lincoln Member across much of Colorado by Kauffman (1967a) and is regarded generally as the basal Lincoln unit in Colorado. Throughout the south-central Western Interior basin, fresh exposures of shale lying above the biostrome are calcareous and locally contain thin calcarenite beds; this suggests that the base of the biostrome can be used widely to mark the Graneros-Greenhorn contact, including the area of the Model anticline. With the contact interpreted in this way, the upper part of the upper shale member of the Graneros Shale is closely comparable in lithology and thickness between Cimarron County and the Model anticline area, except that a greater amount of shale with calcareous silt grains and more numerous thin calcisiltite beds occur near the top of the member in the Oklahoma section. As redefined herein, the Lincoln Member around the Model anticline is 9.8 m (32 ft) thick.

In the Lincoln Member of the Greenhorn Formation a significant change of facies occurs between Cimarron County and the Model anticline. West-northwest from Oklahoma the number of calcarenite beds decreases by nearly 50 percent (although persistent calcarenites become thicker bedded), and chalky limestone beds are rarer or absent in the Model anticline area. In the Cimarron County section, the predominant rock type of the Lincoln is grayish-orange and yellowish-brown shaly chalk and chalky shale. These grade west-northwestward into a section consisting of yellow-brown, grayish-brown, and tan-gray calcareous shales, in part containing calcisilt, with lesser amounts of chalky shale. The trend toward decrease in total carbonate content continues westward to the southern frontal ranges of Colorado, where gray calcareous shale dominates much of the Lincoln section.

At locality 2 in Cimarron County, 8 m (26 ft) of exposed Hartland overlie a calculated 27.5-m-thick (90-ft) covered interval, below which are 5.2 m (17 ft) of exposed Lincoln strata. Thus the total thickness of the Lincoln-Hartland sequence in this area appears to be 40 m (133 ft). The recessive, deeply weathered nature of the covered interval may indicate that it represents dominantly the Hartland Member. This is considerably greater than the thickness of the Lincoln-Hartland interval around the Model anticline area or along the southern Colorado frontal ranges (18.5 and 32 m [61 and 105 ft], respectively). An error in the covered-interval measurement, possibly owing to minor structural complications in the covered interval rather than sedimentologic phenomena, probably accounts for this discrepancy. The Hartland Member also manifests change of facies from Cimarron County to the Model anticline, grading from orange, yellowish-gray, and light olive-gray shaly chalk to tan, yellowish-brown, and light gray to light olive-gray calcareous shale and chalky shale. Southeastern Colorado sections of the Hartland contain fewer thin calcarenites or harder microcrystalline limestones and in Colorado generally the member contains fewer bentonites and chalky limestones.

The Bridge Creek Member thins westward and northwestward from Cimarron County: the exposed part is 13 m (43 ft) thick in Oklahoma, and the equivalent interval is 10.7 m (35 ft) thick on the Model anticline; the same interval averages 10.4 m (34 ft) in thickness along the southern frontal ranges of Colorado. As in Kansas, the Bridge Creek Member extends stratigraphically higher in the Model anticline section than in the Cimarron County section, the latter having suffered considerable erosion of chalky limestone beds that characterize the upper Bridge Creek elsewhere. Because of this, previous remarks on comparative thicknesses of the member are based on equivalent stratigraphic intervals, not on total Bridge Creek thickness.

Lithologically, the principal ledge-forming limestone beds are very similar in the Bridge Creek Member of Oklahoma and Colorado, although in Cimarron County the limestone beds are usually more chalky and in Colorado some of the limestone beds are harder, darker in color, and more argillaceous than those of Cimarron County or Kansas. The greatest lithologic difference is in the softer shaly layers interbedded with the limestones. These are speckled, light olive-gray shaly chalks in Oklahoma, but grade in Colorado into light gray chalky shales and calcareous shales. Certain marker beds can be traced between Oklahoma and Colorado with little change; these beds are the basis for a system of precise lithologic correlation between the two areas, as follows: (1) In Oklahoma, as throughout southeastern and south-central Colorado, a nodular-weathering limestone bed, the second unit (unit 78, sec. 2) of the Bridge Creek Limestone lies 0.33 m (1.1 ft) above the base of the member and contains abundant *Sciponoceras gracile* (Shumard). Bass and others (1947) used this bed to define the base of the Bridge Creek on the Model anticline, but our subsequent investigation suggests that a thin hard bed of limestone just below it serves more logically as the base of the member. (2) A prominent 15-cm-thick (0.5-ft) bentonite lying about 2m (6 ft) above the base of the Bridge Creek Member in Oklahoma is believed to correlate with a 11-cm-thick (0.35-ft) bentonite lying 1.2 m (4 ft) above the base of the Model anticline section; no other thick bentonite occurs in the lower part of either section. (3) Five to seven limestone beds containing abundant *Mytiloides mytiloides* (Mantell) occur in the upper metre of the member in both the Cimarron and Las Animas County,

Colorado, sections; within this suite several beds can be matched exactly.

Comparison with South-Central Colorado. Comparisons as detailed as those with southeastern Colorado are not warranted for the more distant outcrops along southern frontal ranges and in Huerfano Park, Colorado. However, some interesting general comparisons can be made. Kauffman's studies of Huerfano Park (1961) and the eastern margin of the southern frontal ranges (1967a) are the main bases for comparison with Oklahoma. Facies changes between the two areas were documented by Kauffman and others (1969).

In parts of south-central Colorado the Graneros Shale rests on blue-gray siliceous shales and siltstones of a Mowry Shale wedge rather than on sandstone of the Dakota Group as a southeastern Colorado and Oklahoma. The lower shale member of the Graneros, which averages 18 m (60 ft) thick in both areas, contains greater amounts of siltstone and sandstone in its lower part along the Front Range, and its upper part is less silty and sandy than it is farther to the southeast. No siltstone or sandstone occurs above the middle of the member. Bentonite, silty calcarenite, and concretion (siderite and limestone) beds are more abundant in the southern Colorado sections. In south-central Colorado the Thatcher Limestone Member is represented by one or two zones of massive gray limestone concretions and lenses that rarely exceed 1 m (3 ft) in thickness or 2 m (6 ft) in length. Commonly, these are calcarenitic and have a cone-in-cone or siderite perimeter. The upper shale member of the Graneros is considerably thicker in south-central Colorado than in the Oklahoma Panhandle (20 versus 7.2 m [65 versus 23.6 ft]), contains less silty and sandy shale, more bentonite beds and concretions, and greater amounts of limonite and gypsum.

The Lincoln Member apparently thickens westward from approximately 5.2 m (17 ft) in Oklahoma, to 9.8 m (32 ft) in Las Animas County, to 12.2 m (40 ft) in Huerfano County, Colorado. Along the eastern margin of the southern frontal ranges, Lincoln shale beds are darker and less chalky than those of Cimarron County and consist predominantly of olive-gray to gray calcareous shale. Furthermore, the Colorado section contains fewer beds of calcarenite; these are thicker than those in Oklahoma and are dominantly very light gray, medium-grained and have black fish bone fragments scattered throughout. The Oklahoma calcarenites are darker gray to yellowish brown. Finally, in south-central Colorado the Lincoln contains more bentonite beds than in Oklahoma.

The Hartland Member manifests similar changes westward, being less chalky, darker gray, and more bentonitic than the same member in Oklahoma. Chalky limestones are fewer and thin calcarenites are more common than in Oklahoma. The member averages 12.2 m (40 ft) in thickness along the southern Colorado frontal ranges; present knowledge suggests that the member may be somewhat thicker in Oklahoma.

Limestones of the Bridge Creek Member are notably darker in color and more argillaceous than those of Cimarron County; chalky limestone beds are rare and usually lie near the top of the member. In south-central Colorado the layers interbedded with limestone are medium to light gray calcareous shale and grade southeastward to shaly chalk that weathers yellowish and brown. The member thins from 13 m (42.8 ft) to 10.4 m (34 ft) between Cimarron County and the Huerfano Park area of south-central Colorado, probably owing to lateral gradation of limestone beds westward into shaly limestone and calcareous shale.

Comparison with North Texas Section

From Dallas and Tarrant Counties, Texas, northward to the Oklahoma border, detailed lithologic and paleontologic studies of the outcropping upper Woodbine and Eagle Ford Formations demonstrate the equivalency in age and environment of this marine succession with the Graneros-Greenhorn-Carlile sequence of the Western Interior. Although distances of 650 to 730 km (400 to 450 mi) without intervening outcrops separate northwestern Oklahoma and this northern Texas area, significant lithologic similarities exist in equivalent Upper Cretaceous rocks. The Eagle Ford Formation contains the strata of greatest significance to our comparative study. The Eagle Ford is exposed along the western edge of the East Texas Embayment in western Dallas and eastern Tarrant Counties in an outcrop that extends northward to the Red River in Grayson and Fannin Counties. Detailed stratigraphic studies are in progress, but stratigraphic nomenclature has not yet reached the level of refinement of equivalent rocks in the Western Interior.

The rock-stratigraphic nomenclature used in the Oklahoma Panhandle and adjacent areas (Fig. 2) cannot be applied in northern Texas, although an overall likeness is evident and some strikingly similar lithic sequences are noted. General similarities between the two areas include a stratigraphically ascending progression from older sandy nearshore deposits to younger, soft calcareous shale, calcarenite, and chalky beds representing an offshore environment. Biostratigraphic correlations suggest that these physically similar sequences are not exactly the same age. This is to be expected in view of the fact that the line of correlation between the two areas lies at a large angle to the nearly north-south depositional strike in the eastern half of the Western Interior Cretaceous seaway. Diachronous transgressive-regressive relationships are involved.

Equivalency in age between the upper Woodbine-Eagle Ford and the Graneros-Greenhorn-Carlile sequence of the Western Interior has long been known through the efforts of Hill (1887), Cragin (1893), Hyatt (1903), Adkins (1933), Moreman (1942), and Stephenson and others (1942). However, direct physical or biostratigraphic comparisons between the two areas have not been attempted previously. Only Stephenson and Reeside (1938), Reeside (1957), and Hattin (1967) have ventured more than passing mention of the Late Cretaceous connection between the Gulf Coast and the Western Interior regions.

Weimer (1960), Hattin (1964), and Kauffman (1967a, 1969a) have documented successfully the Late Cretaceous marine cycles of deposition in the Western Interior in Colorado, Kansas, and adjacent areas. Major cyclothems are recognizable also in the Upper Cretaceous strata of Texas and are now being studied (Powell, in prep.); a first attempt at comparison with Western Interior cyclothems is outlined below.

Whereas the Dakota Group occupies the earliest position in the Greenhorn Marine Cyclothem of the Western Interior, the Woodbine Formation of northern Texas, lying unconformably above the Comanchean Series, represents the earliest phase of a major marine transgression in Texas (Fig. 4). The Dakota Group is dominated by massive fluvial, deltaic, and shoreline sand bodies, especially in southern Colorado, but includes also dark shales containing carbonaceous silt, variegated mudstones, and brown to gray lenticular sandstones deposited in fluvial, deltaic, bay, and back-bar environments. Woodbine facies and depositional environments are identical, with the only significant variation being local increase or decrease in sand content. Woodbine and upper Dakota faunas have numerous taxa in common (Hattin, 1967; Siemers, 1971) especially among the bivalves, gastropods, and foraminifers.

The Dakota Group and the Woodbine Formation occupy the same relative position in what is apparently the same marine cyclothem (Greenhorn Cyclothem), but they are of overlapping rather than of equal age. Most of the Dakota is Albian in age; only the upper part of the group is of Early Cenomanian age. The Woodbine

Formation is of Middle to early Late Cenomanian age. This indicates that between the Western Interior and north Texas areas the early part of the Greenhorn Cyclothem is offset chronologically, perhaps by as much as half a stage. This offset may reflect the time lag in transgression from the Western Interior region to northern Texas, a distance of 480 to 810 km (300 to 500 mi). Biostratigraphic data suggest that during the latter part of this cycle, the similar depositional patterns of Cimarron County (and areas to the south) and northern Texas were more nearly equivalent in time (Fig. 3).

The Graneros Shale equivalent in northern Texas (lower part of Eagle Ford Formation) includes soft, medium to dark gray, gypsiferous silty clay shales with dark brown very fine grained sandstones that grade upward into less quartzose, more calcareous, bentonite-bearing claystones. Sedimentary structures and fossils indicate marginal to shallow-marine (inner-shelf) conditions of deposition and represent the transition between the paralic Woodbine strata below and the offshore, main body of the Eagle Ford above. This is essentially the same as the depositional environment of the Graneros Shale across much of the south-central Western Interior (Kauffman and others 1969) and surrounding areas (compare Hattin, 1967, p. 587). In spite of intervening distances and the probable differences in source-area contributions, this general environment produced closely similar lithofacies in Texas and Oklahoma, although similar rocks in the two areas are not precisely the same age. The depositional succession and regional relationships involved here represent a classic transgressive sequence predictable within the sedimentary model for the basin (Kauffman, 1969a, Fig. 4).

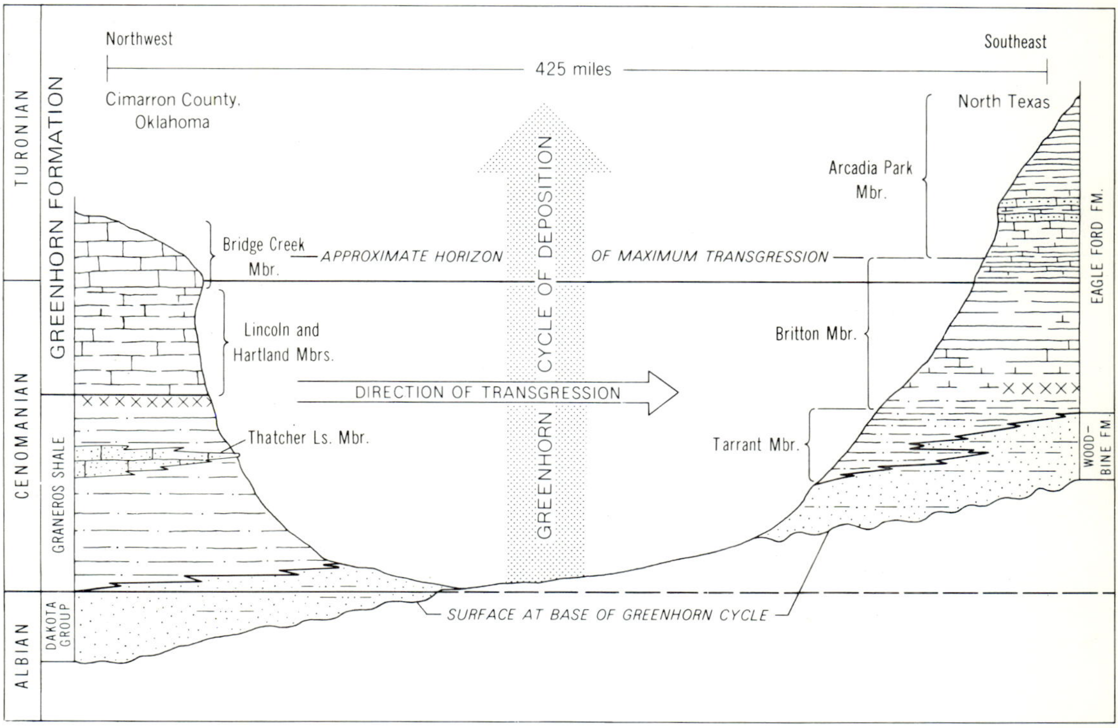

Figure 4. Transgressive part of the Greenhorn Cyclothem showing relationship between stratigraphic sections in Cimarron County, Oklahoma, and north Texas. Schematic; not to scale. See Figure 3 for lithologic symbols.

In Grayson and Fannin Counties, Texas, the Woodbine Formation includes an upper shaly member (Templeton Member of Bergquist, 1949) that is correlative in part with the Tarrant Member of the Eagle Ford (Moreman, 1933) in Dallas and Tarrant Counties. These two members plus the lower 2.5 to 3 m (8 to 10 ft) of the Britton Member of the Eagle Ford are correlative with the lower shale member, the Thatcher Limestone Member, and the upper shale member of the Graneros Shale in Cimarron County, Oklahoma, and adjacent areas (zones of *Inoceramus? bellvuensis* through *Ostrea beloiti*).

The Templeton Member is about 26 m (85 ft) thick and consists predominantly of dark olive-gray gypsiferous silty shale and muddy calcareous sandstone. The Tarrant Member includes from 0.3 to 5.5 m (1 to 18 ft) of yellowish-brown to olive-gray silty shale containing beds of slabby calcareous siltstone and very fine grained sandstone (Powell, 1968). These two members are generally more sandy than the lower shale member of the Graneros in Cimarron County.

In northern Texas, rocks correlative with the upper shale member of the Graneros are included in the lower part of the Britton Member of the Eagle Ford. These rocks are only 3.6 to 4.6 m (12 to 15 ft) thick and include dark brown to gray, weakly calcareous silty shale containing thin (12-cm [0.4-ft]) lenses of brown, very fine grained sandstone and a few very thin (0.6- to 3-cm [0.02- to 0.1-ft]) stringers of calcarenite in the upper part. About 1.5 m (5 ft) above this interval is a bed 27.5 to 30.5 cm (0.9 to 1.0 ft) thick of nearly white bentonite that weathers orange. It is biostratigraphically at the level of the widespread X bentonite marker bed of the Western Interior region, including the Cimarron County section (Fig. 3). The *Ostrea beloiti* biostrome commonly found in association with the X bentonite is not known from any Texas exposure, but in Tarrant County *O. beloiti* does occur as isolated individuals in the shale just below the thick bentonite bed. At this level the upper shale member of the Graneros and lowermost Britton are coeval and represent the approximate time at which lithologic phases of the depositional cycle became essentially synchronous.

The middle part of the Britton Member of the Eagle Ford Formation is correlative with the Lincoln-Hartland interval in Oklahoma. Just above the thick bentonite bed, in Tarrant County, the Britton contains dark olive-gray chalky shale, numerous seams of bentonite, and scattered, thin laminae of calcarenite. Collectively these beds reach as much as 24.5 m (80 ft) in thickness and grade upward into an interval 13.7 to 15.3 m (45 to 50 ft) thick that is characterized by numerous tough, moderately resistant, bluish-gray chalk beds ranging in thickness from 6 to 18 cm (0.2 to 0.6 ft). Interbedded with the chalk are soft, dark gray, chalky shale layers and bentonite seams as much as 24 cm (0.8 ft) thick. The total interval (38 to 40 m [125 to 130 ft]) is within the same age range as the Lincoln-Hartland beds in the Oklahoma Panhandle. The two sections also have a great similarity in appearance and lithology.

In northern Texas the upper part of the Britton Member of the Eagle Ford is more than 61 m (200 ft) thick. It is composed of dark gray clay shale with minor amounts of quartz silt and large numbers of small, flattened, reddish-brown clay-ironstone nodules and light gray septarian limestone concretions. No lithic counterpart of this interval exists in Cimarron County unless it is much thinner and concealed in the covered interval represented by unit 48 of section 2 (Fig. 3). Biostratigraphic evidence suggests that nearly all of the upper Britton interval is equivalent to the *Sciponoceras gracile* Zone, and the upper 4.5 to 6 m (15 to 20 ft) are equivalent to the *Mytiloides opalensis* and possibly part of the *M. mytiloides* Zones in the lower and middle parts of the Bridge Creek Member (Fig. 4) exposed in Cimarron County. The upper Britton shale apparently represents a localized addition from eastern source areas of terrigenous material that overwhelmed and diluted the otherwise highly calcareous interval that is correlated with the upper Hartland(?) and lower Bridge Creek Members.

Rocks correlative with the upper part of the Bridge Creek Member, not preserved in Cimarron County because of erosion, are included in the Arcadia Park Member of the Eagle Ford Formation (Fig. 4). The basal unit of the Arcadia Park is from 3 to 9.5 m (10 to 30 ft) thick and is composed mainly of dark gray calcareous clay shale; hard, dark brownish-gray shaly chalk, and hard beds 8 to 20 cm (0.2 to 0.6 ft) thick of medium grayish-orange *Inoceramus*-prism and foraminiferal calcarenite (inoceramite of Hattin, 1962, p. 41). This unit contains *Mytiloides mytiloides* and *M "latus"* and is therefore also equivalent to the upper part of the Pfeifer Member of the Greenhorn Formation and the lowermost part of the Fairport Member of the Carlile Shale in Kansas (Hattin, 1962, 1975) and adjacent areas of Colorado. The main body of the Arcadia Park consists of dark olive-gray, soft clay shale that weathers into flakes. The shale is strikingly similar to that in the Blue Hill Shale Member of the Carlile Shale in Kansas and Colorado, and the two are equivalent biostratigraphically.

In summary, the upper Dakota-Graneros-Greenhorn-Carlile sequence of Cimarron County manifests considerable lithologic similarity and is homotaxial with the Woodbine-Eagle Ford sequence of northern Texas. The transgressive succession of depositional environments from fluvial-deltaic to offshore shelf is, with minor exceptions, practically the same in Oklahoma and northern Texas. Southeastward, from near the basin center (including the Oklahoma Panhandle), elements of the Greenhorn Marine Cyclothem can be traced to north Texas, near the eastern margin of the basin. Because of the time lag owing to transgression, rocks in the lower part of the cyclothem are younger in a southeastward direction (Fig. 4).

Transgression over such large distances must have traversed irregular topography, which would result in local or regional variations in lithostratigraphic detail. By the time basal paralic environments of the Greenhorn marine cycle had progressed eastward beyond the north Texas area, most of the basin was characterized by offshore environments. During the height of transgression, carbonate deposits with approximate time equivalency were deposited in Oklahoma and northern Texas (Fig. 4).

Summary of Facies Relationships of the Oklahoma Section

In general, the section of Cretaceous sedimentary deposits studied in the Oklahoma Panhandle is closely similar in lithology, succession of facies, and age (as defined biostratigraphically) to rocks of the Greenhorn Cyclothem transgression exposed in central and western Kansas and southeastern Colorado. Somewhat less similar, but nevertheless generally age-equivalent lithostratigraphic units occur also along the eastern flank of the Wet Mountains and in Huerfano Park, south-central Colorado. Southeast (Oklahoma) to northwest (Colorado) facies changes primarily involve decrease in chalk content, increase in argillaceous material, and general decrease in sand-sized quartz and calcite grains in the supra-Dakota sequence. The Graneros Shale and the Lincoln Member of the Greenhorn Formation thicken significantly west-northwestward from Oklahoma and thin slightly northeastward into Kansas. The Hartland and Bridge Creek Members thin to the west-northwest; the Hartland apparently thins but the Bridge Creek does not change thickness in a north-northeastward direction into western Kansas. Marked time-transgression of the lower part of the section occurs between Oklahoma and northern Texas and between Oklahoma and central Kansas. Facies changes of stratigraphically equivalent rocks are sufficient to warrant use of separate stratigraphic nomenclature for the Texas sections. The homotaxial equivalents of clay-rich beds of the Oklahoma section are notably more sandy and silty in Texas, and the carbonate-rich part of the Oklahoma section changes southeastward into a more clay-rich, thicker sequence.

Widespread, distinctive marker beds have played an important role in confirming lithologic and (or) chronologic equivalence in several parts of the section. The Thatcher Limestone Member of the Graneros Shale is widespread in southern Colorado, but this marker is not recognized in northern Texas or east of the westernmost part of Kansas. The X bentonite is relatively thick in all areas studied and lies amongst or adjacent to beds containing scattered valves or biostromes of *Ostrea beloiti*. Within the Bridge Creek Limestone Member of Oklahoma are a bentonite seam and at least two limestone beds that can be traced with confidence to northern Kansas and central Colorado. However, regional correlation and facies relationships have been established largely on biostratigraphic evidence.

BIOSTRATIGRAPHY

Upper Cretaceous exposures of northwestern Cimarron County include rocks which are assignable to the Cenomanian through Lower Turonian Stages of the European sequence, as these stages are defined in the Western Interior (Cobban and Reeside, 1952; Kauffman, 1967a, 1967b). They contain faunas that are a mixture of Western Interior and Gulf Coast taxa and also include several cosmopolitan or intercontinentally distributed elements. Species typical of the Western Interior, plus certain more cosmopolitan elements, dominate both the macrofaunas and microfaunas. The Cimarron County section is therefore critical to the integration of highly refined systems of biostratigraphic zonation that represent the two most extensive Cretaceous biogeographic divisions of the North American Province—the Gulf and Atlantic Coast Subprovince and the Western Interior endemic center (Kauffman, 1973, Figs. 1, 2). Although this has not been attempted in detail previously, some general relationships have been documented by Stephenson (1952, p. 20, 21), Stephenson and Reeside (1938, p. 1629–1638), and Cobban and Reeside (1952, chart 10b). Cosmopolitan and intercontinental taxa, which include many well-known European species, are dominantly planktonic foraminifera, ammonites, and inoceramid bivalves; these are the basis for regional correlation and application of European stage nomenclature to the Upper Cretaceous sequence of the Oklahoma Panhandle.

In general, the Upper Cretaceous strata of Cimarron County yield sufficient well-preserved fossils to allow accurate dating within the integrated biostratigraphic-geochronologic system outlined by Kauffman (1970), detailed biostratigraphic zonation (zones of 0.17- to 0.1-m.y. duration), and correlation with equivalent Gulf Coast and Western Interior sections.

Only possible worm and arthropod burrows are known from marginal-marine and shallow-water Upper Cretaceous sandstones in the underlying upper part of the Dakota Group. The silty lower shale member of the Graneros shale is equally barren; not even arenaceous foraminifera have been recovered. The Thatcher Limestone Member of the Graneros contains few specifically recognizable macrofossils; these are mostly shark teeth and inoceramid and ostreid bivalves retaining the shell. The bivalves are useful in dating and correlating this member with the type Thatcher and with exposures to the west and southwest. The upper shale member of the Graneros yields abundant, well-preserved calcareous and arenaceous foraminifers (dominantly planktonic species in the upper part) and scattered macrofossils. The macrofauna consists almost entirely of bivalves preserved as moderately compressed internal or composite molds (McAlester, 1962) on bedding surfaces. These are well-enough preserved to permit specific identification and correlation.

The Greenhorn Formation is richly fossiliferous throughout; in fresh exposures most bedding surfaces of the Lincoln Member, Hartland Member and Bridge Creek Limestone Member yield identifiable macrofossils consisting predominantly of mollusks. Many of these are internal, external, or composite molds showing a moderate to high degree of morphologic detail. Bivalves commonly retain some shell material, and most structures are well preserved; ammonites and gastropods retain gross structures, but sutures and fine ornament are rarely preserved. Specimens in hard limestone beds are not crushed; those in chalk or shale are greatly flattened parallel to the bedding, but without lateral distortion. Planktonic foraminifera dominate the microfossil assemblages of chalk, marlstone, and calcareous shale beds of the Greenhorn; preservation is generally poor because specimens were crushed and deformed during compaction. Nevertheless, specific identification is generally possible.

Macrofossils in the Cimarron County section are dominantly mollusks, in particular ammonites belonging to the Acanthoceratidae, Baculitidae, Desmoceratidae, Hamitidae, Scaphitidae, and Anisoceratidae, and bivalves of the families Inoceramidae, Pteriidae, Ostreidae, and Pectinidae. Other bivalves, gastropods, scaphopods, possible ahermatypic corals, patellid brachiopods, and determinate vertebrate remains (other than fish scales and teeth) are rare. Hermatypic corals, many gastropod families, echinoderms, bryozoa, articulate brachiopods, and arthropods (except for burrows) are absent here just as throughout much of the Western Interior Cretaceous strata. Faunal elements that are best preserved in the Cimarron County section are those that are of primary biostratigraphic importance throughout the Western Interior and Gulf Coast sequences (ammonites, inoceramids, oysters, and planktonic foraminifera). From our collections, a detailed local biostratigraphic zonation is possible (Fig. 5), and the Cimarron County section can be correlated faunally and lithologically with Cretaceous sections in adjacent areas.

Western Interior

The biostratigraphic standard now in use throughout the Western Interior is that of Cobban and Reeside (1952) as subsequently refined by detailed lineage studies of ammonites (Cobban, 1952, 1953, 1958a, 1958b, 1961, 1962a, 1962b, 1964, 1969; Cobban and Scott, 1964, 1972; Reeside and Cobban, 1960), bivalves (Kauffman, 1965, 1966a, 1967b); gastropods (Sohl, 1967), foraminifera (Eicher, 1960, 1965, 1966; Eicher and Worstell, 1970; Kent, 1967, 1968; Mello, 1969, 1971), and nannoplankton (Trexler, 1967). This standard is now being revised by a group of specialists who are utilizing *all* major elements of the Cretaceous biota (Kauffman and Cobban, eds., in prep.), through application of concepts and methodology described by Kauffman (1970). Projected refinement of this major revision is approximately equivalent to the biostratigraphic revision proposed here for Cimarron County in the more fossiliferous upper part of the section (Hartland and Bridge Creek Members of the Greenhorn Formation).

The Oklahoma section contains sufficient well-preserved macrofossils to permit biostratigraphic correlation with the present Western Interior standard (see Fig. 5). In addition, a local zonation can be constructed from our data (Fig. 5), utilizing concepts elaborated by Kauffman (1970). The local zonation further refines the Western Interior standard zonation in this area. The duration of local assemblage

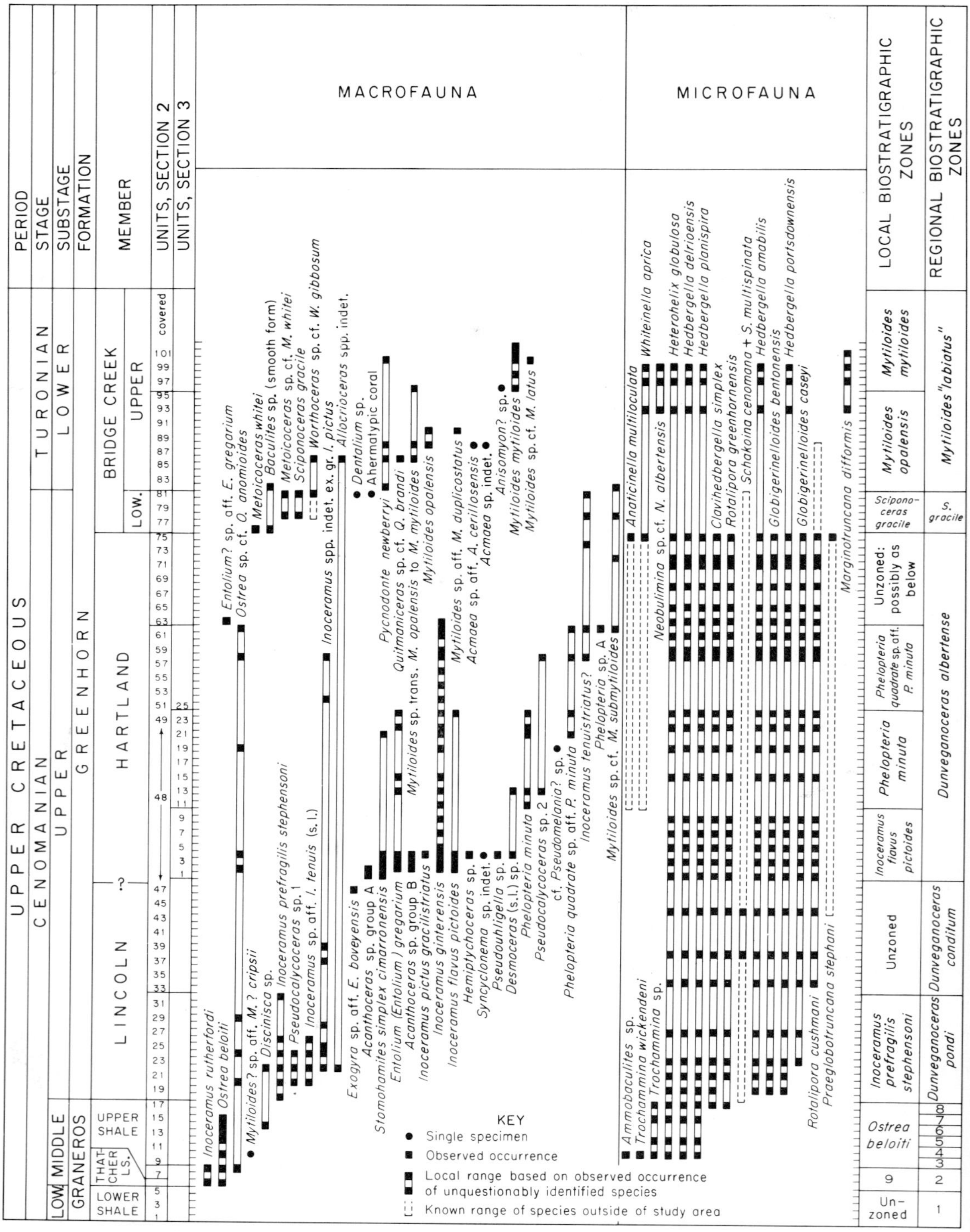

Figure 5. Biostratigraphic plot of all faunal elements studied from Cenomanian and Lower Turonian strata of Cimarron County, Oklahoma. Proposed local biostratigraphic zones are based on assemblages of all well-defined taxa and are designated by key species in each assemblage; regional (Western Interior) biostratigraphic zones from Cobban and Reeside (1952) and Kauffman (1971a). Explanation in text. Numbered regional biostratigraphic zones (lower right) as follows: 1, *Inoceramus bellvuensis–I. dunveganensis*; 2, *Calycoceras* (*Conlinoceras*) *gilberti–C. leonense–Borrisjakoceras compressum–Turrilites acutus* Concurrent Range Zone; 3, *Calycoceras* (*Conlinoceras*) *gilberti–Acanthoceras* cf. *A. plicatile* Concurrent Range Zone; 4, *Acanthoceras granerosense*; 5, *Acanthoceras muldoonense*; 6, *Acanthoceras alvaradoense*(?); 7, *Acanthoceras amphibolum*; 8, *Plesiacanthoceras wyomingense*. 9, Local zone of *Inoceramus rutherfordi*.

zones, as determined from a geochronologic standard (Kauffman, 1970, Figs. 8 through 10; 1977) introduced into Oklahoma through correlation of bentonites and widespread faunal zones, is 0.5 to 1.0 m.y. per zone in the sparsely fossiliferous sequence from the Graneros Shale through the Lincoln Member of the Greenhorn Formation and 0.17 to 0.25 m.y. per zone in the richly fossiliferous Hartland Member and Bridge Creek Limestone Member of the Greenhorn. No previous zonation has been established for the Oklahoma Upper Cretaceous strata.

Gulf Coast

In Texas, strata of the Eagle Ford Formation that are equivalent to the Graneros Shale and Greenhorn Formation of Cimarron County yield similar faunas, although the Texas strata also contain distinctive, warmer water (Subtropical) southern elements. Moreman (1927, 1942) established the first comprehensive biostratigraphic zonation for the Eagle Ford based predominantly on ammonites and *Inoceramus*. This has been modified somewhat by later studies (Powell, 1963a, 1963b, 1965, 1968, work in progress). Pessagno (1967) summarized the foraminiferal biostratigraphy for this sequence; he drew partially from earlier systematic treatments.

Powell is currently restudying the stratigraphy and faunas of the Eagle Ford Formation and constructing a revised biostratigraphic zonation comparable in refinement to that of the Western Interior Cretaceous strata. Foraminiferal data are being integrated with data for the dominant mollusks (ammonites, bivalves) following the format being used for biostratigraphic revision of the Western Interior Cretaceous faunas (Kauffman, 1970). Our correlation of the Oklahoma and Texas sections utilizes Powell's refined, unpublished zonation and is thus more precise than would have been possible with published data. Integration of Western Interior radiometry (Kauffman, 1970; Obradovich and Cobban, 1973; Williams and Baadsgaard, 1973; Kauffman, 1977) into Powell's zonation through correlation of equivalent faunal zones and bentonites suggests that, as currently defined, the Eagle Ford zones have an average span of 0.5 m.y.

Correlation of the Cimarron County Cretaceous Section

The recorded biostratigraphic ranges of all recognizable taxa found in Cimarron County during our study are shown on Figure 5. Utilizing the system of zonation described by Kauffman (1970), these data have been used to establish a local biostratigraphic zonation (right side of Fig. 5) that is compared with the Western Interior standard and Eagle Ford zonation currently in use. It should be noted that certain ammonite indices used to identify the Western Interior standard zones were not recorded in the Oklahoma section (for example, *Acanthoceras* spp., *Dunveganoceras* spp.); correlation with these zones (Fig. 5) was accomplished through mutually occurring species of oysters and inoceramids.

Dakota Group. In Cimarron County the upper part of the Dakota Group yielded no recognizable fossils other than burrows in the area studied. Age assignment of this thin sequence, lying above the thick-bedded Dakota sandstones, to the earliest part of the Cenomanian is based on the fact that the outcrops in northwestern Oklahoma are virtually surrounded by areas (New Mexico, Colorado, Kansas) where similar uppermost Dakota strata are of Early or Middle Cenomanian age. The upper Dakota of Colorado and New Mexico is of Early Cenomanian age (Kauffman, 1967a; Kauffman and others, 1969; Scott, 1970); in central Kansas the upper Dakota contains a fauna (Hattin, 1967; Siemers, 1971) that compares with the latest Early? and Middle Cenomanian faunas of the Texas Woodbine

Formation (Stephenson, 1952). Dating and correlation of the Dakota is based in all cases predominantly on Bivalvia. The geographic location of Cimarron County suggests strongly that the upper Dakota there is of Early Cenomanian age, and probably equivalent to the lower part of an assemblage zone containing *Pachydiscus brazoensis* Böse, *Inoceramus? dunveganensis* McLearn, *I.? athabaskensis* McLearn, and a late evolutionary variant of *I.? bellvuensis* Reeside.

Graneros Shale. On the basis of its position between the probable Lower Cenomanian beds of the upper Dakota Group and the Middle Cenomanian Thatcher Limestone Member of the Graneros Shale (*Calycoceras tarrantense* (Adkins) Zone), the lower shale member of the Graneros is assumed to be Early Cenomanian in age and equivalent to the upper part of a Western Interior assemblage zone containing *Inoceramus? dunveganensis, I.? athabaskensis,* and *I.? bellvuensis* (late form).

The Thatcher Limestone Member in Oklahoma marks the first appearance of *Inoceramus rutherfordi* Warren. Elsewhere in the Western Interior this species first appears in the latest Early to early Middle Cenomanian *Calycoceras tarrantense* Assemblage Zone of Cobban (in Kauffman and Cobban, eds., in prep.). Significantly, similar *Calycoceras* occur in the type Thatcher so it would seem that the Thatcher of Oklahoma lies in the same zone. Kauffman and others (1969) reported a seemingly distinct ammonite fauna (*Acanthoceras wintoni* fauna) from the higher of two zones of Thatcher-like limestone lenses and concretions that they interpreted as being slightly younger (earliest Middle Cenomanian) than that of the type Thatcher. This suggests that the Thatcher might therefore be slightly time-transgressive toward the east. Subsequent collections have shown that the *A. wintoni* fauna also contains *Calycoceras tarrantense* and other ammonites typical of this zone, which suggests that the Oklahoma, southeast Colorado, and south-central Colorado exposures of the Thatcher are generally of equivalent age (late Early to early Middle Cenomanian). The Thatcher Member occupies the lower half of the local zone of *Inoceramus rutherfordi* in Oklahoma (Fig. 4).

The upper shale member of the Graneros Shale embraces the entire Middle Cenomanian Substage and contains the first abundant foraminifera in the sequence (Fig. 4). Benthonic species predominate except at the top of the member where planktonic forms become dominant and remain so through the remainder of the exposed section. Planktonic forms are more numerous and diverse in the upper Graneros of Oklahoma than in equivalent beds of Kansas and Colorado. The arenaceous taxa are more useful in precise correlation, however, because all of the planktonic species are long-ranging. *Hedbergella delrioensis, H. planispira,* and *Heterohelix globulosa* first appear in the upper Graneros in Oklahoma and range upward to the upper Bridge Creek Member of the Greenhorn Formation. These species also first occur in the upper Graneros of Kansas and the foothills of the Colorado frontal ranges and thus provide a general basis for correlation. The planktonic species *Rotalipora greenhornensis* and *Clavihedbergella simplex* also first appear in the uppermost Graneros beds in Oklahoma (Fig. 5), but only the former has been reported from the upper Graneros of Colorado or Kansas (Eicher and Worstell, 1970). The appearance of rotaliporids in the uppermost Graneros manifests the environmental changes that brought about the deposition of Greenhorn carbonates and possibly indicates that there is little or no hiatus between the Graneros and Greenhorn in Oklahoma.

Ostrea beloiti Logan is the only abundant macrofossil in the upper shale member of the Oklahoma Graneros. Sparse but biostratigraphically important species are *Inoceramus rutherfordi* Warren (lower part) and *Mytiloides?* sp. aff. *M.? crippsi* (Mantell) (upper two-thirds of the member). Diagnostic acanthoceratid ammonites, used elsewhere to divide the upper Graneros into three widely traceable zones (Fig. 5), were not observed in Cimarron County. The lower two beds of the member are assigned to the upper half of the latest Early to early Middle Cenomanian local zone of *I. rutherfordi* (=*Acanthoceras wintoni* Zone of Kauffman and others, 1969; now = upper part of *Calycoceras tarrantense* Zone). The upper half of the member is assigned to the Middle Cenomanian *Ostrea beloiti* Assemblage Zone recognized at equivalent stratigraphic levels in Kansas by Hattin (1965a) and in Colorado by Kauffman (1967a). *Mytiloides? crippsi* (Mantell) is rare in America, but is one of the most common Middle Cenomanian inoceramids of Europe.

Greenhorn Formation. Ammonites and inoceramid bivalves are the principal basis for zonation of the Greenhorn Formation in the Western Interior. Species of *Dunveganoceras* and members of the *Inoceramus pictus* Sowerby lineage traditionally constitute the zonal indices of the Lincoln and Hartland Members, but only the inoceramids were found in the Cimarron County section. Other ammonites and bivalves, however, permit us to establish a new zonal division of the Hartland. The lower part of the Bridge Creek Member is generally marked by a diversity of ammonites, most of which occur in the Oklahoma section, comprising the *Sciponoceras gracile* (Shumard) Assemblage Zone. The middle and upper parts of the Bridge Creek are placed traditionally in the zone of *Mytiloides labiatus* (Schlotheim, s. l.). This zone has been subdivided by Cobban and Scott on the basis of ammonites (1972, p. 33). An alternative subdivision of this zone, based on the members of the *M. labiatus* lineage and pycnodontid oysters, is a principal biostratigraphic contribution of our paper. These new bivalve zones are widely traceable in the Western Interior. Foraminifera, especially planktonic forms, are abundant throughout the Greenhorn and are an important accessory group in the biostratigraphy of this sequence.

In the southern part of the Western Interior (Cobban and Reeside, 1952, chart 10b) the Lincoln Member generally includes the zone of *Dunveganoceras pondi* Haas. Subdivision into a lower *D. pondi* Haas-*Metoicoceras praecox* Haas Zone and an upper *D. conditum* Haas-*M. muelleri* Cobban Zone has been suggested by Cobban (in Kauffman and Cobban, eds., in prep.). Hattin (1968, p. 1084) extended the zone of *Acanthoceras wyomingense* Reagan to include the basal beds of the Lincoln in southwestern Kansas. None of these key ammonites were observed in the Lincoln of Oklahoma; however, the member did yield *Pseudocalycoceras?* sp. (generic range Late Cenomanian to early Early Turonian). In southwestern Kansas, *Pseudocalycoceras* occurs with *Acanthoceras wyomingense* Reagan in the lower part of the Lincoln Member. Inoceramid bivalves constitute the best means of correlating the Lincoln of Oklahoma with other Western Interior sections. *Inoceramus prefragilis stephensoni* Kauffman and Powell, n. subsp., and *I.* sp. aff. *I. tenuis* Mantell occur with other bivalves, rare serpulids, gastropods, and brachiopods in the lower half of the member; macrofossils are rare in the upper half. The subspecies *I. prefragilis stephensoni* has been recognized also at this stratigraphic level in Colorado and New Mexico where it is associated with *Dunveganoceras pondi* and thus permits correlation of the lower Lincoln in these areas and establishes an early Late Cenomanian age for the member. *Inoceramus prefragilis stephensoni* is the only common fossil restricted to the lower Lincoln in Oklahoma and has been chosen as the local zonal index in this area. Lack of diagnostic fossils prohibits zonation or precise correlation of the upper Lincoln Member above unit 32 of section 2 (Fig. 3).

Foraminifera of the Lincoln Member are predominantly planktonic species (listed in Fig. 5), as noted also by Morrow (1934) for Kansas and by Eicher and Worstell (1970) for Kansas and Colorado. Calcareous benthonic species are rare in these beds, and arenaceous forms are absent; this suggests unfavorable ecological conditions at the sediment-water interface during deposition of the Lincoln. Nine species of foraminifera make their first appearance at the base of the Lincoln in Oklahoma and strongly complement inoceramids in establishing the base of the local *I. prefragilis stephensoni* Zone. However, none of the Lincoln foraminifera is restricted to that member (Fig. 5).

Foraminifer range zones are of little use in the biostratigraphic division of the Lincoln Member, but changes in relative abundance, size, and form of some species may be useful for future subzonation if comparison with other areas establishes their regional significance. For example, units 19 and 21 of section 2 (Fig. 3) contain huge numbers of medium-sized (0.2 to 0.35 mm) *Hedbergella delrioensis*, abundant *H. planispira*, and moderate numbers of the other planktonic forms. Unit 23 of section 2 also contains huge numbers of robust *H. delrioensis* and moderately abundant *H. portsdownensis*, but the numbers of *H. planispira* are vastly reduced. Unit 34 is the lowest level at which rotaliporids are abundant. Large rotaliporids (0.7 to 0.8 mm) continue to be common elements of the fauna through unit 44. For some distance above unit 44, rotaliporids are rare, but they reappear in abundance and greater diversity in the Hartland Member (unit 65, sec. 2).

In summary, the Lincoln Member of Oklahoma seems to represent only a single recognizable local assemblage zone, that of *Inoceramus prefragilis stephensoni*, which is restricted to the lower half of the member and apparently correlates with the widespread zone of *Dunveganoceras pondi*.

In the Oklahoma section the Hartland Member is richly fossiliferous throughout. Although the primary Western Interior zonal index for the member, *Dunveganoceras albertense* (Warren), has not yet been found here, other ammonites, numerous bivalves including cosmopolitan inoceramids, and a rich planktonic foraminifer assemblage make possible the biostratigraphic subdivision of the Hartland into three local assemblage zones (Fig. 5). Each of these zones is represented in equivalent beds in southern Colorado or northern New Mexico. The precise lateral stratigraphic relationships of these zones are now under study; they are potentially useful regionally.

The lowest local assemblage zone contains the exclusive known range of *Inoceramus flavus pictoides* Sornay in the Western Interior and the entire local ranges of *Pseudouhligella* sp., "*Acanthoceras*" sp. (groups A and B), and *I. pictus gracilistriatus*, n. subsp. (this subspecies ranges to the end of Cenomanian time in Europe). This local zone is named the *Inoceramus flavus pictoides* Assemblage Zone. *Inoceramus ginterensis* Pergament, *Entolium* (*Entolium*) *gregarium* Kauffman and Powell, n. sp., and *Stomohamites simplex* (d'Orbigny) first appear and are common in this zone, but range upward into the overlying zone. The predominantly planktonic assemblage of foraminifera within this zone is the same as that found in the upper Lincoln Member and ranges upward throughout the Hartland (Fig. 5). This is clearly a Late Cenomanian assemblage and confirms correlation with the Hartland Member elsewhere in the Western Interior. More precise correlation of this zone with the lower part of the Hartland in Colorado, southwestern Kansas, or New Mexico is suggested but not yet substantiated by inoceramid and ammonite evidence.

The Hartland microfauna compares closely with that described by Morrow (1934) from the Hartland of Kansas and includes important, restricted Late Cenomanian forms such as *Rotalipora cushmani*. Ammonites in the lower part of the Hartland also indicate a Late Cenomanian age. The forms called "*Acanthoceras*" s. l. may belong to as yet unnamed generic groups. Several such "nameless genera" are known to occur in the Upper Cenomanian strata in North America. Poor preservation prevents assignment of our immature specimens to such genera as *Tarrantoceras*, *Pseudocalycoceras*, *Acanthoceras*, or *Dunveganoceras*, the young of which can be of very similar appearance.

Equivalent rocks in areas outside of the Western Interior—such as the far western Gulf of Mexico region in northern Coahuila, Mexico, and Trans-Pecos, Texas—offer the only known comparative fauna published to data (Powell, 1965; Maxwell and others, 1967, p. 65). In Coahuila, concretionary limestone near the base of the Boquillas Formation contains an acanthoceratid fauna (Powell, 1965, p. 521) including *Tarrantoceras* sp., *Desmoceras* (*Pseudouhligella*), *Tragodesmoceroides*, *Pseudocalycoceras*, *Idiohamites*, *Hypoturrilites?*, a nearly smooth, small baculite, and *Inoceramus* of the *I. pictus* group. Collectively, these forms indicate a Late Cenomanian age and compare generally with the lower Hartland ammonite-inoceramid assemblage. *Stomohamites simplex* (d'Orbigny) occurs at several places in the lower and middle Hartland of Colorado as well as in the Eagle Ford Formation of north Texas. Because of diachrony of the Lincoln-Hartland contact, this form occurs only in the Lincoln of central Kansas. *Inoceramus flavus pictoides* and *I. pictus gracilistriatus* occur in undifferentiated Upper Cenomanian beds of Europe. Both species are rare in the Western Interior; to date the former is known only from lower Hartland beds in Oklahoma and Colorado. The lowest local assemblage zone of the Hartland Member extends from units 1 through 8 of section 3 (App. 1; Fig. 3).

The next higher local assemblage zone includes units 9 through 24 of section 3 and equivalent units of section 2 (see App. 1 and Fig. 3) and spans the lower-middle part of the Hartland Member. The zone is named *Phelopteria minuta* Assemblage Zone after a new, moderately common pteriid, the range of which is restricted to this part of the Hartland in Oklahoma, south-central Colorado, and locally in northeastern New Mexico. *Inoceramus ginterensis*, *Entolium* (*Entolium*) *gregarium*, *Stomohamites simplex*, and *Desmoceras* sp., s. l., range commonly into, but not definitely above, this zone from the underlying zone of *I. flavus pictoides*. Planktonic foraminifera of this zone are abundant, but all recorded species occur elsewhere in the Hartland Member of Oklahoma, Kansas, and Colorado. In total, the assemblage of this zone is similar to that of the lower half of the Hartland Member in Colorado and also generally similar to that of Upper Cenomanian deposits from Mexico to Colorado and Kansas.

In the upper middle part of the Hartland Member (units 51 through 62, sec. 2), *Phelopteria minuta* Kauffman and Powell, n. sp., is replaced by a descendant here designated *Phelopteria* quadrate species aff. *P. minuta*. The two forms overlap in unit 24 of section 3. This form is differentiated from *P. minuta* by its more quadrate outline (see Pl. 8, figs. 4, 6 through 8), but insufficient well-preserved specimens are available for formal description. A similar evolutionary sequence in *Phelopteria* has been noted by Kauffman in south-central Colorado. Although other fossils are relatively rare in this interval (see Fig. 5), a distinct local range zone is proposed tentatively for this easily recognized form. The stratigraphic range of *Pseudocalycoceras* terminates at the top of the *P.* quadrate species aff. *P. minuta* Range Zone in Oklahoma, but this genus commonly ranges higher in the Cretaceous section elsewhere in the Western Interior. The lower part of the range of *Inoceramus tenuistriatus* lies in the uppermost part of the *P.* quadrate species aff. *P. minuta* Range Zone. The limited assemblage of this local zone is of Late Cenomanian age and correlates with the upper middle part of the Hartland in south-central Colorado.

The uppermost part of the Hartland Member in Cimarron County (units 63 through 69, sec. 2; App. 1 and Fig. 3) contains few recognizable fossils and cannot be assigned with confidence to any local zone, although it seems to be most closely related to, and probably a part of, the *Phelopteria* quadrate species aff. *P. minuta* Range Zone. This part of the member contains a suite of typical Late Cenomanian planktonic foraminifera (Fig. 5) and poorly preserved fragments of *Inoceramus* belonging to both the underlying *I. ginterensis* and overlying *I. tenuistriatus* species groups. In addition it contains, at the top, the exclusive local range of the foraminifera *Praeglobotruncana stepheni* and *Anaticinella multiloculata* (regionally known throughout all or most of the Hartland; see Fig. 5), and the lowest local occurrence of *Whiteinella aprica*, also a long-ranging form found lower in the Hartland elsewhere. Because this interval lacks a diverse assemblage of ammonites, including *Sciponoceras gracile*, these beds are assigned tentatively to the upper part of the *Dunveganoceras albertense* Zone of the Western Interior succession.

The Bridge Creek Limestone Member of the Greenhown Formation, to the extent that it is exposed in Cimarron County, is divisible into three regionally widespread assemblage zones. The lowest of these, the *Sciponoceras gracile* (Shumard) Assemblage Zone, is one of the most distinctive and widespread biostratigraphic units in the American Cretaceous section. It includes units 76 through 81 in section 2 (see App. 1 and Fig. 3). Elsewhere in the Western Interior the zone is characterized by a considerable diversity of ammonites, inoceramids of the *I. pictus* Sowerby lineage, and an unusually diverse assemblage of other mollusks. Many of the ammonites and inoceramids of this zone occur also in Europe. Throughout North America, and to a lesser degree in Europe, this zone is considered to be the highest level in the Cenomanian; many Europeans still consider it basal Turonian. The assemblage recorded in the lower Bridge Creek of Oklahoma is less diverse than normal but contains typical highest Cenomanian elements (as defined in North America): the ammonites *Sciponoceras gracile* (Shumard), *Metoicoceras whitei* Hyatt, and species of *Worthoceras; Inoceramus tenuistriatus?* Nagao and Matsumoto, one of the youngest members of the *I. pictus* Sowerby lineage; and *Mytiloides* sp. aff. *M. submytiloides* (Seitz), the earliest ancestor of the *M. labiatus* (Schlotheim) lineage. *Sciponoceras gracile* and *M. whitei* are restricted to this zone over most of the Western Interior; *Worthoceras* and *M.* sp. aff. *M. submytiloides* range slightly higher into the basal Turonian, and *I. tenuistriatus* occurs also slightly lower in the section in the upper one-fourth of the Hartland Member. The *Worthoceras,* though specifically indeterminate, compares most closely with species found in the *S. gracile* Zone of Texas, Kansas, and Colorado.

The *S. gracile* Zone occurs exclusively in the lower two or three limestone beds of the Bridge Creek Member in northeastern New Mexico and Colorado. In western Kansas the zone also lies in the lower part of the Bridge Creek Member, but in central Kansas the zone occurs within the lower one-half of the Hartland Member. This apparent downward shift in position results from the disappearance to the east of many limestone beds of the lower part of the Bridge Creek such that in central Kansas the lower Bridge Creek equivalent is classed as a part of the Hartland. The *S. gracile* Zone fauna of Oklahoma is known from an interval at least 60 m (200 ft) thick that lies in the middle part of the Eagle Ford Formation of north Texas. In addition to *S. gracile, M. whitei,* and *Worthoceras* (*W. vermiculum* (Shumard) in Texas), the Texas equivalent of the zone also contains *Metoicoceras acceleratum* Hyatt, *Kanabiceras septemseriatum* (Cragin), *Pseudocalycoceras* spp., *Allocrioceras* spp., *Proplacenticeras* spp., *Scaphites brittonensis* Moreman, *Scaphites* sp., and *Borissiakoceras* sp. Similar elements are found in the Kansas and Colorado equivalents of the *S. gracile* Zone. Identical inoceramids and gastropods occur in all areas, and conspecific oysters occur in Colorado and Oklahoma sections.

In Oklahoma the base of the *S. gracile* Zone marks the extinction level of a great majority of the planktonic foraminifera that characterize the Lincoln and Hartland Members. Bridge Creek foraminifers are very abundant but considerably less diverse than those of older strata. This may be due in part to the fact that well-preserved material could be obtained only from units 93, 97, and 99 of section 2, the other beds being either too hard (limestone) or too deeply weathered. No foraminifera have been recovered from the *S. gracile* Zone in Oklahoma, although it is assumed that *Hedbergella portsdownensis, H. amabilis, H. planispira, H. delrioensis, Heterohelix globulosa,* and *Whiteinella aprica*—which occur below (Fig. 5) and above this interval—could be collected from that zone were the chalky shales less weathered.

Cobban and Reeside (1952, chart 10b) originally assigned the *S. gracile* Zone to the base of the Turonian Stage. However, the close relationship of the ammonites, inoceramids, and other mollusks to older (Cenomanian) forms and the magnitude of the faunal (evolutionary) break between the *S. gracile* Zone and the overlying zone of *Mytiloides* "*labiatus*" (in the traditional sense; not as here revised) have now been recognized. Therefore, the zone is here classed as the uppermost zone of the Cenomanian Stage (for example, Kauffman, 1970, Fig. 4; Kauffman and others, 1969).

Restudy of the lineage of *Mytiloides labiatus* (Schlotheim), traditionally used as the zonal index of the middle and upper parts of the Bridge Creek Limestone Member in the Western Interior (for example, Cobban and Reeside, 1952, chart 10b; Hattin, 1965b; Kauffman, 1967b, 1970, Fig. 4; Kauffman and others, 1969, Fig. 3), was first undertaken seriously by Seitz (1934 [1935]), who demonstrated that Schlotheim's type specimen of *M. labiatus* was distinct from specimens being assigned to that species in Europe and North America; the common form generally called *Inoceramus* (*Mytiloides*) *labiatus* is, in reality, *Mytiloides mytiloides* (Mantell). The species is abundant in the upper part of the Bridge Creek Limestone Member throughout the southern and central Western Interior, including Oklahoma. Detailed population analysis of the *M. mytiloides* lineage (Kauffman, 1970, Fig. 4) has revealed further that sufficient morphological changes take place within the middle and upper parts of the Bridge Creek to permit systematic division of the lineage into at least two species upon which two widely applicable biostratigraphic zones can be based. These zones are (ascending) the *Mytiloides opalensis* (Böse) Assemblage Zone and *M. mytiloides* (Mantell) Assemblage Zone; both are defined formally in Oklahoma for the first time, although the zones have already been traced widely in the Western Interior region.

The *Mytiloides opalensis* Zone, which corresponds to the *Watinoceras coloradoense* Zone of Cobban and Scott (1972, p. 33), marks the basal part of the Turonian Stage and in Oklahoma extends from unit 82 through unit 95 in section 2. The zone is further characterized by abundance of *Pycnodonte newberryi;* this oyster ranges into the lower part of the overlying *M. mytiloides* Zone but is more diagnostic of the lowest Turonian strata throughout the Western Interior. In the uppermost bed of the zone, *M. opalensis* co-occurs with its descendant, *M. mytiloides.* Ammonites are sparse and poorly preserved in this zone and include *Allocrioceras?* sp., *Worthoceras?* sp., *Baculites* sp. (small smooth form), and *Quitmaniceras* sp. cf. *Q. brandi* Powell. The last is of importance because *Q. brandi* was described by Powell (1963a) from the lowest Turonian (Salmurian) strata of northeastern Chihuahua, Mexico, in a fauna that contains faunal elements of both latest Cenomanian and earliest Turonian affinities.

Collectively, the biostratigraphic evidence suggests that the *M. opalensis* Zone may be considered as the basal unit of the Turonian Stage in the Western Interior. In many parts of Colorado and New Mexico, inoceramids are rare or absent in this zone, ammonites very rare, and the zone can be recognized mainly as an interval yielding relatively abundant *Pycnodonte newberryi* and common limonite and marcasite(?) nodules that lies between the richly fossiliferous zones of *S. gracile* and *Mytiloides mytiloides*.

The highest zone represented in the preserved part of the Oklahoma Bridge Creek is that of *Mytiloides mytiloides* (Mantell), which is equivalent to at least the upper one-half of the *M. labiatus* Zone of previous American authors (for example, Cobban and Reeside, 1952). This bivalve is abundant throughout the interval of units 96 through 102 in section 2 where the species occurs typically with Early Turonian species of foraminifera (Fig. 5). The *M. mytiloides* Zone is approximately equivalent to the *Mammites nodosoides* Zone of Cobban and Scott (1972, p. 33).

Oklahoma foraminifera obtained from the *M. mytiloides* Zone include some forms that range upward from Middle and Upper Cenomanian strata (Fig. 5) as well as two newly introduced species. The last are *Marginotruncana difformis* (Gandolfi) and *Neobulimina* sp. cf. *N. albertensis* (Stelck and Wall). The double-keeled species of *Marginotruncana* occurs in the Eagle Ford Formation of north Texas in association with *Mytiloides* "*labiatus*" (probably *M. mytiloides*). Pessagno (1967) gave a reasonably complete synthesis of the occurrence of these species and concluded that the base of the Turonian Stage can be defined as coinciding with the earliest occurrence of double-keeled *Marginotruncana*. *Mytiloides mytiloides* is considered to be of Early Turonian age throughout the world and is perhaps one of the most cosmopolitan of the Inoceramidae. *Mytiloides* "*latus*" (in the sense of Hattin, 1962), a more rounded, evolutionary derivative of *M. mytiloides* that marks a distinctive zone in the overlying Fairport Member of the Carlile Shale first appears toward the top of the *M. mytiloides* Zone. *Pycnodonte newberryi* is sparse in the lower part of the *M. mytiloides* Zone in Oklahoma (Fig. 5).

Biostratigraphic correlation of the Upper Cretaceous section of Oklahoma with the Western Interior standard and application of European stage nomenclature are readily accomplished in the Greenhorn Formation. Data gathered from the Oklahoma section suggest that further subdivision of the Western Interior standard may be possible once the faunal detail of this interval is more widely known. In the Oklahoma section the upper part of the Dakota Group and the Graneros Shale are sparsely fossiliferous; only generalized correlation, combining faunal and lithologic data, can be effected for these units.

PALEOENVIRONMENTS

Marine paleoenvironments represented by the Cretaceous rocks of Cimarron County and the surrounding region are decipherable through study of sedimentary characteristics and faunal assemblages. General stratigraphic and environmental relationships between strata of southern Colorado and Oklahoma have already been discussed (Kauffman and others, 1969), but continuing research on Oklahoma material and the addition of information from Kansas and Texas now permit more detailed regional interpretation.

Paleoenvironments of the Cenomanian-Turonian sequence in Kansas and Colorado have been defined previously by Hattin (1964, 1965a, 1965b, 1967, 1975) and Kauffman (1967a, 1969a). Kauffman (1967a, 1969a) described the composition and paleoecology of 36 distinct, molluscan-dominated paleoassemblages (predominantly Bivalvia) that characterize the Cretaceous rocks of the Western Interior and provide a primary source of paleoenvironmental data. Many of these occur in the Cimarron

County section and are discussed subsequently.

Although the limited collections from the Oklahoma section do not constitute statistically significant samples, the techniques used in the interpretation and definition of the fossil assemblages in the Oklahoma Cretaceous section are generally the same as those applied by Kauffman (1967a). Individual collections were tabulated as to physical and biological parameters regardless of the level from which they came and were analyzed for (1) consistency of biotic association and (2) consistency of lithologic association. Assemblages recorded frequently in the collections were deemed significant, and those characteristic of a specific lithotype were considered environmentally diagnostic. All but one of the assemblages defined here were already known to be distributed widely in equivalent rocks elsewhere in the Western Interior (Kauffman, 1967a, 1969a).

Seven assemblages can be distinguished in the fossiliferous part of the Oklahoma section. No attempt was made to define ichnofossil associations. All but one assemblage are dominated by one or two taxa or morphotypes, and all are of limited diversity. Restrictive environmental factors, such as water and substrate chemistry (especially salinity), substrate firmness and composition, turbidity, circulation, and food supply are probably factors responsible for these low-diversity assemblages.

Fossil Assemblages

Assemblage Dominated by Robust, Byssate Inoceramids and Small Ostreids. (Assemblage V (part), Kauffman, 1967a; assemblage 9 (part), Kauffman, 1969a.) In Oklahoma, *Inoceramus rutherfordi* and *Ostrea beloiti*, together with abundant fish scales, bone and tooth fragments, and sparse fragments of other mollusks, compose this assemblage in units 6 through 8 of the Graneros Shale (sec. 2; Fig. 3). This assemblage occurs where these units are silty and calcarenitic adjacent to the Thatcher Limestone Member and also in well-cemented calcarenites of the Thatcher (lithologies 4a and 5, Kauffman, 1969a). This parallels the occurrence of the assemblage elsewhere in the Western Interior Cretaceous marine section.

The unusual concentration of fish, *Inoceramus*, and *Ostrea* debris that forms much of the calcarenite strongly suggests low rates of terrigenous detrital sedimentation in a high-energy regime where organic remains were broken, concentrated, and winnowed. At various Colorado and New Mexico exposures, and to a limited degree in Oklahoma, small-scale cross-bedding, scour surfaces, and irregular lenticular bedding occurs in the Thatcher Limestone Member; this suggests shallow inner-shelf environments within reach of main longshore and tidal currents, and possibly large waves (that is, generally less than 50-ft depth). Many modern analogues exist.

Organisms characteristic of this assemblage are well adapted for a shallow, active-water environment. *Inoceramus rutherfordi* has a moderately thick inequivalve shell, further strengthened by strong sulci and folds, and possessing strongly developed pedalbyssal musculature that indicates strong byssal attachment to the substrate. Shell fragments or even calcarenitic sediment would have formed a suitable substrate for firm anchorage; the shell lay on the inflated left valve, propped above the sediment-water interface by the umbo, fold(s), and posterior auricle to prevent sediment fouling of the apertures and to enhance feeding and respiration.

Ostrea beloiti, whereas relatively small and thin-shelled compared to Cretaceous oysters from nearshore, high-energy environments (for example, *Lopha*, *Crassostrea*, and certain *Exogyra*) is nevertheless well adapted for shallow-water and active-current environments. The slipper-shaped left valves were firmly attached to a hard substrate, to other shell fragments, or to a partially cemented calcarenite surface; they coiled slightly upward with the inhalent-exhalent apertures elevated above the substrate and the flat upper valve tilted so that settling sediment did not accumulate over it. Firm attachment and upward growth of the species in crowded biostromes adapted *O. beloiti* for life in agitated water. Modern small oysters of this type, and in this gregarious growth mode, are known primarily in waters from littoral to 15-m (50-ft) depths.

These two forms not only dominate this assemblage, but occur also as less abundant elements in more diverse nearshore assemblages elsewhere in the Western Interior (Kauffman, 1969a). Their occurrence in Cimarron County without diverse associates may indicate that certain environmental factors—possibly substrate mobility and constant current scouring—may have been deleterious to less firmly attached organisms and (or) that the shells of other species were destroyed during constant reworking of the sediment. Interestingly, a low-diversity ammonite assemblage dominated by one or two genera of large, coarsely ornate forms with or without associated baculites, scaphites, and placentiform ammonites characterizes nearshore, high-energy deposits elsewhere in the Western Interior but has not been recorded in the Oklahoma section.

Assemblage Dominated by Large, Flat, Thin-shelled Inoceramids. (Assemblage X of Kauffman, 1967a; assemblage 12 of Kauffman, 1969a.) Thin, evenly bedded, silty clay shales of the upper Graneros Shale generally imply slow, occasionally interrupted sedimentation in quiet (restricted circulation) to gently agitated waters of midshelf depth (30 to 120 m [100 to 400 ft]); the shales deposited in the shallowest part of this range are characterized by more silt, thin siltstones, and fewer bentonites with lateral continuity (for example, lower shale member of Graneros). Vertical compression and gradational boundaries of burrows within this shale further imply a soft, water-saturated substrate. Such conditions imposed severe restrictions on the types and adaptations of associated benthonic organisms. A common Cretaceous assemblage associated with presumed soft clay and carbonate substrates in the Western Interior includes thin-shelled, nonbyssate to weakly byssate inoceramids of low convexity, together with their boring or encrusting epibionts on the up-facing valve. Such inoceramids seem to have been adapted for life on more fluid substrates. *Mytiloides* sp. aff. *M.? crippsi* of the upper Graneros Shale is such a form; in life it lay recumbent on the left valve and "floated" free as a thin plate or very weakly attached by byssal threads to shell fragments. Among common epibiont associates in the upper Graneros Shale (units 10 through 15, sec. 2) where this assemblage is recognized in Oklahoma are small encrusting *Ostrea beloiti* and the inarticulate brachiopod *Disciniscа.* Slow sedimentation rates, low current velocities, and soft, water-saturated, substrata are suggested by the assemblage.

Assemblage Dominated by Robust, Inequivalve, Nonbyssate to Slightly Byssate Inoceramids. (Assemblage V (part), Kauffman, 1967a; assemblage 9 (part), Kauffman, 1969a.) Detailed morphologic analysis of inoceramid shells has demonstrated conclusively that the moderately robust inequivalve inoceramids, lumped into a single widely occurring *Inoceramus*-dominated paleoassemblage by Kauffman (1967a, 1969a), occur in two distinct habitat groups. One group of strongly byssate forms (for example, *I. rutherfordi*) occurs mainly in strata indicating high- to moderate-energy environments; robust inoceramids occurring in offshore, quieter-water environments were nonbyssate to weakly byssate and lived mostly free on the substrate, reclining on the larger left valve, propped by the umbonal ridge, folds, and the posterior auricle so that the apertures were above the sediment-water interface. Epibionts and borings into the shell are distributed in such a manner on many specimens as to support this view of life orientation. Assemblages dominated by the latter type (namely, *I. pictus gracilistriatus, I. ginterensis, I. flavus pictoides, I. tenuistriatus?, I.* sp. aff. *tenuis*, and *I. prefragilis stephensoni*) are abundant in shale, marlstone, offshore calcarenite, and limestone facies of the Lincoln, Hartland, and Bridge Creek Members of the Greenhorn Formation in Cimarron County. The weight, shape, and moderate size of these valves, plus their thicker shells, seemingly prohibited them from living on very soft, water-saturated substrates without sinking. Byssal attachment appears to have been too weak for them to have withstood highly active current and wave action in shallow, unprotected environments.

In Oklahoma the latter group can be divided tentatively into four subassemblages on the basis of commonly associated organisms; the divisions grade into one another. These subassemblages are listed below:

1. Inoceramids associated only with sparse epibionts on or boring into the shell (oyster spat, *Discinisca?*, boring sponges, worms, barnacles). This subassemblage occurs in many beds of the Lincoln Member and sparsely in the lower and middle parts of the Hartland Member in Oklahoma. Foul bottom conditions, where only the most tolerant organisms (inoceramids) could survive, are implied. Limitation of associated organisms to the surface of *Inoceramus* shells suggests chemically deleterious conditions at the sediment-water interface. The paucity of benthonic foraminifera through much of the Oklahoma Greenhorn supports this interpretation.
2. Inoceramids associated with abundant, gregarious *Ostrea* and small *Exogyra* as encrusting and bottom-dwelling forms. Firm substrate and normal circulation are implied. Ostreids, like inoceramids, were apparently also highly tolerant bivalves during the Cretaceous and lived in environments that excluded less tolerant organisms. This subassemblage characterizes certain lower and middle Lincoln strata in Oklahoma.
3. Inoceramids associated with moderately abundant *Phelopteria* spp. In modern seas most pteriids are strongly attached bysally to a firm substrate that is ordinarily elevated above the sediment-water interface. Abundance of pteriids in this subassemblage is due to colonization of exposed surfaces of inoceramids and (or), more likely, to erect vegetation rooted in the substrate. This is a favorite habitat, along with other sessile organisms such as octacorals, for living pteriids. Ammonites (*Pseudocalycoceras, Stomohamites*) are sparsely associated; benthonic foraminifera are rare. As compared with the preceding subassemblages, the greater diversity might indicate more nearly normal marine, offshore, quiet to gently agitated water conditions during deposition of the middle and upper parts of the Hartland Member, where this assemblage is common.
4. Inoceramids associated with abundant small, very thin shelled pectinids, *Entolium* or *Syncyclonema* (lower Hartland Member). The dominant habitat of these so-called "paper pectens," on the basis of comparison with modern counterparts, is moderately soft to firm substrate in quiet, commonly deep offshore waters, where they can occur in great abundance locally. Normal chemistry of the substrate surface is inferred from abundance of these benthonic bivalves. The ecology of the pectinids is further discussed below.

Assemblage Dominated by Small, Thin-shelled Pectinids. (Assemblage 24, Kauffman, 1969a, p. 240.) Abundant small, thin-shelled pectinids occur today mostly in offshore, quiet-water environments at depths exceeding 30 m (100 ft). Here they occur in association with moderately soft to firm substrates under normal, open-marine conditions. Many species range into shallower water, where they occur as secondary elements of other communities; but in deeper water, pectinids dominate many marine assemblages, especially where they occur in clusters. Thorson (1957, p. 461–534), for example, gave data on North Atlantic pectinid-dominated communities. The "paper pectens" (for example, Propeamusiidae, Amusiidae) have thin shells especially well adapted for life in deeper water muds. The shell is light enough to rest freely on a soft substrate or even to "float" on soupy, water-saturated materials but is too thin to withstand strong currents. The rounded, nearly equilateral shape with reduced auricles and smooth to nearly smooth disc surface gives maximum mechanical swimming efficiency and sensory perception (Kauffman, 1969b)—necessary attributes for survival in poorly lighted, exposed habitats.

The form and sediment association of certain Cretaceous pectinids from the thinly and evenly bedded shaly chalks of the lower and middle parts of the Hartland Member (namely, *Entolium* and *Syncyclonema*) are equivalent to that of modern "paper pectens," and they are considered here to have similar environmental implications. This assemblage occurs in two forms: (1) with shells occurring on numerous laminae, scattered through several millimetres of section and dominating an assemblage that may also include *Inoceramus* and *Phelopteria*—the "normal" living occurrence—and (2) with shells occurring in great numbers on single laminae, almost to the exclusion of other organisms and possibly traceable to New Mexico and Colorado as marker beds. This situation is thought to reflect mass mortality of clusters, especially of *Entolium*, possibly through changes in water chemistry or rapid burial.

Assemblage Dominated by Subequivalve, Slightly to Moderately Biconvex, Nonbyssate to Weakly Byssate Inoceramids. (Assemblage W, Kauffman, 1967a; assemblage 11, Kauffman, 1969a.) Species of the mytiliform *Mytiloides* spp. (*M. submytiloides, M. mytiloides, M. opalensis, M. latus*) dominate this assemblage, almost to the exclusion of other organisms in some cases. The assemblage is associated with carbonate shale and mudstone of the upper Hartland Member (rarely) and especially with limestone of the Bridge Creek Member in Oklahoma. Associates are gryphaeoid *Pycnodonte* (oysters), the heteromorph ammonite *Allocrioceras*, and patelliform gastropods such as *Acmaea* and *Anisomyon* spp. The oysters and gastropods imply a relatively firm substrate, tightly packed or even partially lithified carbonate mud. Photic conditions are implied because the gastropods were probably algal grazers. However, the thin, delicate shells of the reclining epifaunal to semi-infaunal inoceramids, which lacked a byssus for attachment (species have no pedalbyssal muscle scars or byssal slit), suggest a habitat lacking strong current action. Kauffman (1969a) proposed for the Bridge Creek an environment lying within the photic zone but below the effective wave base and swept by only gentle currents.

Offshore Assemblage Dominated by Few Ammonite Taxa. (Assemblage T of Kauffman, 1967a; assemblage 36, 1969a.) In Oklahoma this assemblage is composed of aberrant heteromorphs, especially small delicate forms like *Allocrioceras* and small baculites (*Sciponoceras* or *Baculites*) along with one or more genera of small- to medium-sized ornate ammonites (*Metoicoceras*). It is restricted here to the *S. gracile* Zone of the lower Bridge Creek Member. Baculites commonly dominate. Colorado and Kansas equivalents of the assemblage seem to have higher diversity among the small ornate forms. Various inoceramids of the *I. pictus* and *M. mytiloides* lineages are commonly associated; these include small- to medium-sized, robust inequivalve forms with probably weak byssal attachment and thin-shelled, recumbent nonbyssate forms. Benthonic foraminifera, oysters, and other mollusks are normally sparse in this assemblage. Most ammonites were nektonic or saltating benthonic organisms, and their presence here suggests normal, open-marine conditions, as do the bivalves and predominantly planktonic foraminifera. The aberrantly coiled *Allocrioceras* seems to have been most successful in the offshore carbonate environments supporting this assemblage and probably was a poor swimmer too delicate to resist strong current action. Quiet-water, open-marine conditions are inferred for this assemblage.

Offshore Assemblage Dominated by Diverse Ammonite Taxa. (Assemblage U, Kauffman, 1967a; assemblage 35, Kauffman, 1969a.) As many as ten ammonite genera are found in this assemblage, which in the Western Interior is associated with carbonate mudstones almost exclusively. A maximum of four genera represent the assemblage in any one collection from the Hartland Member of Oklahoma, possibly owing to limited exposure in the section we studied. The ammonites include *Allocrioceras*?, *Worthoceras*, *Quitmaniceras*, *Metoicoceras*, *Pseudocalycoceras*, *Stomohamites*, *Desmoceras*, *Pseudouhligella*, and smooth baculitids, all in association with the sparse *Inoceramus*, *Phelopteria*, *Entolium*, and small oysters. This assemblage, including the benthonic species, seems to represent normal open-marine conditions and certainly higher productivity than the related, but much less diverse, restricted outer-shelf ammonite assemblage. Both assemblages occur in the *S. gracile* Zone elsewhere in the Western Interior.

Paleoenvironmental Interpretation: Physical and Biological Evidence

The Oklahoma section described in this paper represents most of the transgressive part of the Greenhorn depositional cycle. The section contains no physical or faunal evidence of any significant erosional or nondepositional hiatus. The succession of paleoenvironments indicated by physical and faunal data is closely similar to that suggested for the same units in Kansas by Hattin (1962, 1964, 1965a, 1967, 1971, 1975) and in the Raton Basin by Kauffman and others (1969); it is generally similar to Kauffman's reconstructions for the southern part of the Western Interior, especially Colorado (1967a, 1969a).

The initial deposits of the Greenhorn Cyclothem are contained in the main body of the Dakota Group of Aptian-Albian age. These were not studied in detail for this paper. Numerous studies (Waage, 1953, 1955, 1961; Plummer and Romary, 1942; Franks and others, 1959; Hattin, 1967; Siemers, 1971; MacKenzie, 1971; and so forth) have demonstrated that the Dakota consists of a complex of fluviatile, deltaic, and marginal-marine deposits. In Oklahoma the upper part of this sequence, forming a gradation with the Graneros Shale, apparently represents the transition from a marginal-marine to an open-marine environment. Brackish interdistributary-bay, tidal-marsh, lagoon, outer-delta, bar, and foreshore sheet sand environments all have been identified in this interval in the southern part of the Western Interior.

The lower shale member of the Graneros Shale is interpreted as inner sublittoral, where shallow-water clays and migrating sand ripples were laid down in turbid waters. Although recognizable macrofossils are uncommon, there is evidence for considerable organic activity including numerous burrows and *Crossopodia* trackways. Collectively, the trace-making organisms probably accounted for much destruction of bedding planes in these beds and produced considerable suspended material at the sediment-water interface that could have hindered development of diverse benthonic biotas.

Hattin and Frey (1969) defined the probable habitat range of *Crossopodia* sp. as extending from nearshore, shallow-water to offshore deeper water marine environments in fine-grained terrigenous and carbonate sand and in mud without strong physicochemical control. This was thus a broadly adaptive organism, but its trackways are preserved most commonly in siltstone and fine-grained sandstone or calcarenite representing moderately current-swept environments. Scattered to concentrated, broken skeletal remains of fish further attest to moderate currents and (or) intense burrowing and scavenging activity in the environment of deposition of the lower Ganeros. General analysis of faunal assemblages in the lower shale member elsewhere in the Western Interior suggested to Kauffman (1967a, 1969a) a nearshore environment with moderate current and wave action at depths probably not exceeding 30 m (100 ft).

A marked decrease in the rate of terrigenous sedimentation, but continuing moderate to strong current activity in shallow, nearshore waters, is reflected by the calcarenitic Thatcher Limestone Member. The highly fragmented nature of invertebrate skeletons, the worn condition of the bone fragments, and the irregular bedding of the Thatcher attest to persistent action of waves and (or) currents and prolonged winnowing of the finest grades of terrigenous detritus. These calcarenites represent deposition in an offshore zone of wave dissipation as suggested in models of epeiric sedimentation proposed by Irwin (1965) and Shaw (1964). The faunal assemblage of the Thatcher in Oklahoma (described above) is interpreted as characteristic of nearshore current-swept bottoms in less than 15 m (50 ft) of water.

Quieter and deeper water environments are indicated generally by the thinly and evenly bedded shales of the upper shale member of the Graneros, which preserve thin but laterally continuous bentonite seams and contain but little terrigenous silt and fine land-derived carbonaceous debris. Thin shell and bone calcarenites near the top of the member are less argillaceous, more continuous, and more evenly bedded than those below. Carbon specks (plant debris(?)) are scattered and very small. Diversity of fossil assemblages (discussed above) is low. Hattin (1965a) suggested that similar beds in Kansas are broadly analogous to offshore, delta bottomset beds. They may also be analogous to midshelf clays being deposited today in Atlantic and Gulf Coast environments beyond the zone of coarse detrital transport and the reach of waves and nearshore current systems (Kauffman, 1967a, 1969a).

Upper Graneros strata and biota suggest quiet water with current activity that was too slight to disturb sediment layering and sedimentation rates that were at times slow enough to enable oysters to establish small biostromes having four generations of *Ostrea beloiti*. Shell and bone calcarenites accumulated when currents occasionally scoured the bottom, winnowing the fines and concentrating scattered skeletal remains. The occurrence of flat, thin-shelled inoceramids, mostly scattered small oysters, few burrows, and low faunal diversity may reflect a generally soft and water-saturated bottom environment. Presence of abundant, though nondiverse, arenaceous foraminiferal assemblages and epifaunal bivalves suggests adequate water circulation. Upward changes of the foraminiferal assemblage indicate that in Oklahoma the marine environment shifted gradually from nearshore conditions to more offshore conditions as the eastward transgression proceeded. During deposition of the upper Graneros, the site became closer to being the outer edge of terrigenous sand and silt sedimentation. Faunal and sedimentologic evidence in this area and also in equivalent beds of the southern Western Interior indicate depths at least below the normal wave base (possibly 15 to 46 m [50 to 150 ft]) at the beginning of deposition of the upper shale member. Strata higher in the member may have been deposited in water depths of between 46 and 120 m (150 and 400 ft).

The thin, evenly bedded chalky shales and shaly chalks that dominate the Lincoln and Hartland Members and the preservation of widely distributed thin bentonites reflect continued quiet-water deposition below the wave base where sediments were affected for the most part only by gentle bottom currents. Upward increase in calcium carbonate and decrease in terrigenous detritus probably reflect continued eastward transgression of the eastern strandline. This continued transgression left the Oklahoma area relatively closer to the depositional center of the basin. For the most part, therefore, the water remained deep, probably more than 60 m (200 ft). Thin lamination in much of the section indicates slow and frequently interrupted sedimentation.

Thin, widely persistent calcarenites consisting largely of planktonic foraminifera in the Hartland Member may indicate periods of virtual cessation of sedimentation other than by organic tests over large areas. Calcarenites in the Lincoln Member indicate occasional increases in current activity that helped to break up skeletal material and winnow the sediments. Fish debris is common in the inoceramites and often highly fragmented; some teeth are worn by current rolling. Such features gradually diminish upward in the section, which indicates progressive deepening above the base of the Lincoln Member.

In the Lincoln and Hartland Members a general increase in the amount of burrow-mottling, caused primarily by worms, is a manifestation of slow, interrupted sedimentation, which enabled populations of burrowing organisms to become widely established (Rhoads, 1970). Faunal assemblages of the Lincoln and Hartland are dominantly composed of open-water planktonic or nektonic organisms (foraminifera, ammonites) or quiet-water benthonic forms. Thin-shelled pectinids are abundant in many beds, as are small, delicate pteriid bivalves that probably were attached to rooted bottom vegetation or similar upright substrata (especially common in the middle and upper Hartland); if vegetation was involved, photic conditions are indicated. The overall environment was one of quiet, offshore water having a depth range of 45 to 90 m (150 to 300 ft). The larger inoceramid bivalves preserved in these beds were generally nonbyssate to slightly byssate, recumbent, free-living forms that could not have withstood strong current activity without being overturned. The general lack of, or very low levels of, benthonic foraminifera in these deposits and the low diversity of the benthonic mollusk assemblage together indicate unfavorable bottom conditions. In all probability this was due to (1) deleterious sediment chemistry for infaunal organisms and (2) a soft, water-saturated substrate upon which could live only specially adapted forms like paper pectens and inoceramids with a robust lower valve and accessory supporting structures (folds, auricles) to keep the feeding margin above the sediment-water interface.

In Oklahoma, deposition of the Bridge Creek Limestone Member occurred near the basin center for the most part beyond the influx limit of large amounts of terrigenous detritus. Macroinvertebrate skeletal debris, coccoliths, and foraminiferal tests are major components of the sediment. Modern deposits of similar aspect are accumulating mostly in outer-shelf and deep-sea areas. Alternation of shaly chalk beds with purer limestone beds may indicate (1) environmental fluctuations that inhibited or stimulated growth of the calcareous plankton or (2) fluctuations in the influx of very fine grained terrigenous detritus (see Hattin, 1971). Burrow-mottling is common in the limestones, but shaly chalks and calcareous shales are thinly and evenly bedded and contain few burrows; this suggests that the limestone beds represent periods of slow sedimentation (Hattin, 1971). The Bridge Creek Limestone Member lacks evidence of strong current or wave action; there are no calcarenites as in the older Lincoln Member or in the Bridge Creek elsewhere in the Western Interior. However, some of the limestones show faint scour contacts and concentrations of burrows on their upper surfaces that indicate periods of reduced or halted sedimentation followed possibly by some cementation and gentle current scour. Fossil assemblages indicate optimum conditions for planktonic and nektonic organisms (coccoliths, foraminifera, ammonites) during deposition of the lower Bridge Creek Member. The upper Bridge Creek is marked by a great decline in ammonites. Benthonic conditions changed also; moderately diverse bivalves and gastropods recorded in the lower Bridge Creek indicate a relatively stable, normal-marine substrate, whereas the upper Bridge Creek contains a highly restricted benthos of thin-shelled recumbent inoceramids and fragile oysters.

Certain aspects of the Bridge Creek fossil assemblages elsewhere in the Western Interior indicate that this sedimentary deposit accumulated in the photic zone. Among these are sparse rudist bivalves in growth position—some species of which are thought to have symbiotic algae, algal borings, possible calcareous algae crusts—and a presumed shallow-water molluscan assemblage including numerous small oysters, a variety of gastropods (including *Turritella*, aporrhaids, fusids, and cerithiids), and a few infaunal bivalves. These fossils suggest that the environment of sedimentation within the Bridge Creek elsewhere may have provided more favorable benthonic habitats generally than was the case in the Oklahoma area.

PART 2

Paleontology

Erle G. Kauffman
J. Dan Powell

INTRODUCTION

The Cimarron County section has yielded a moderately diverse fauna of considerable biostratigraphic and paleobiogeographic significance. The fauna includes several new taxa as well as some forms that have not been reported previously from the Western Interior basin. The systematic treatment of the biota, exclusive of ostracodes and microflora, is presented in Part 2 of this paper.

The Greenhorn fauna contains a number of previously undescribed bivalve species and the first occurrences in Western Interior Cretaceous strata of certain Gulf Coast, Western European, and (least commonly) Indo-Pacific species of ammonites, bivalves, and (or) foraminifera. These first occurrences are of considerable biostratigraphic importance because they allow long-range correlations with other faunal provinces and realms.

Typical Western Interior faunas, which dominate the Cimarron County assemblages, represent warm- to mid-temperate Cretaceous climatic zones. However, most of the taxa that have been previously described but are here reported from the Western Interior for the first time are either typical subtropical to warm-temperate Gulf Coast forms or species more common in the equivalent climatic zones of Western Europe. This is to be expected because Oklahoma lies at the boundary between the two principal biogeographic divisions of the Cretaceous north-temperate realm in North America—the Gulf and Atlantic Coast Subprovince (subtropical to warm-temperate) and the Western Interior endemic center (warm- to mid-temperate), as defined by Kauffman (1973). Oklahoma also lies at the southern entrance to the Western Interior seaway, along routes that were probably taken by certain European molluscan stocks that migrated into North America during the world-wide Cenomanian transgression (Kauffman, 1973, p. 368, 379, 380). Cimarron County is thus a pivotal area for the integration of regionally discrete biostratigraphic systems and a principal area where biogeographic boundary studies might be carried out. For these reasons, detailed documentation of the Cimarron County Cretaceous fauna is of considerable importance.

Systematic practices employed here are by necessity variable. Where populations are available, as for *Inoceramus ginterensis* Pergament and *Entolium* (*Entolium*) *gregarium*, n. sp., biometric data are presented and population variation is discussed. Figure 1 shows measured features and abbreviations used in the text and tables. Other descriptions are necessarily based on limited material that is well-enough

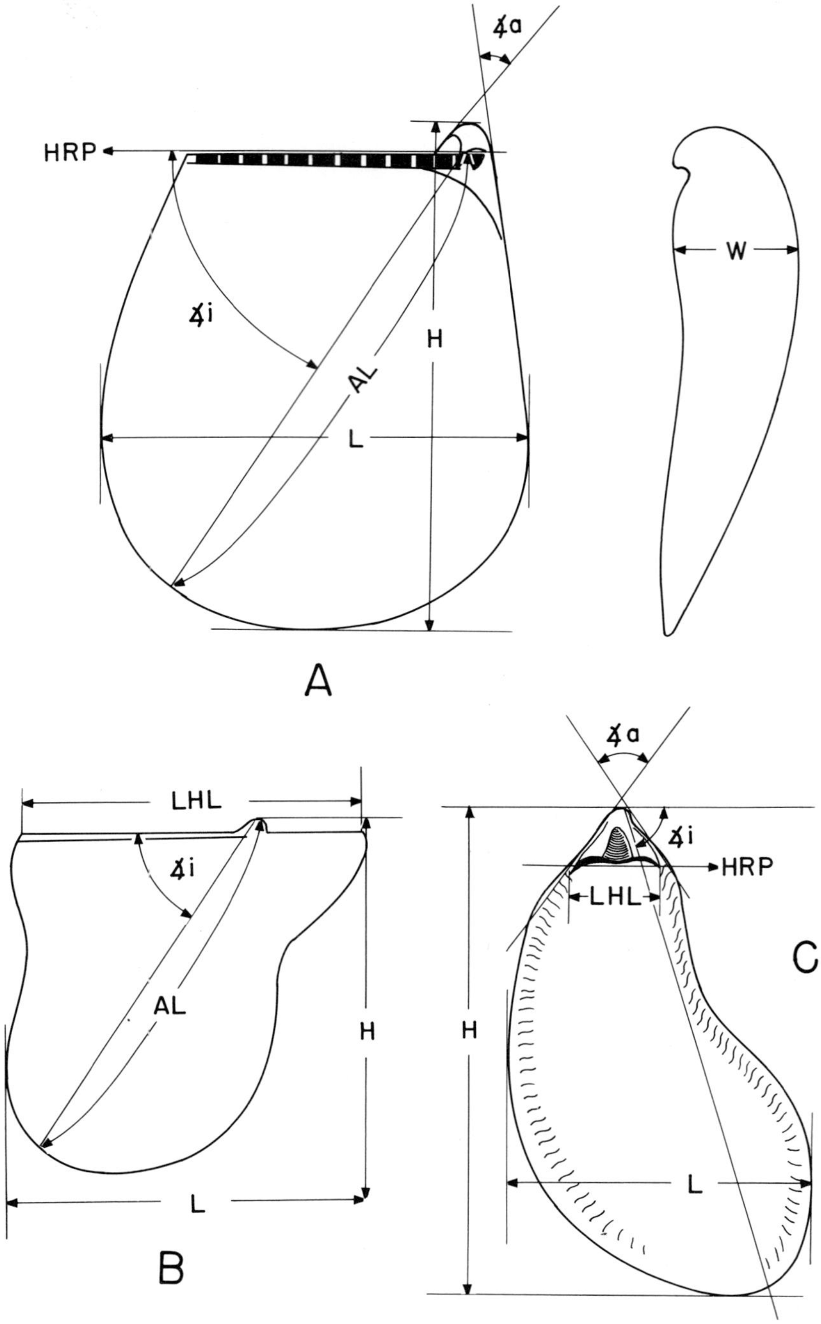

Figure 1. Measured features on Cretaceous bivalves from northwestern Oklahoma, showing exact position of each parameter. Diagram A is schematic after *Inoceramus tenuistriatus*? Nagao and Matsumoto; diagram B is schematic after *Phelopteria minuta*, n. sp.; diagram C is schematic after *Ostrea beloiti* Logan. Key to measured parameters: *a* = apical angle; *AL* = length to any feature as measured along the shell axis (axial length), for example, to any ruga or for the purpose of counting growth lines or rugae per unit length; *H* = height of entire shell; *HRP* = horizontal reference plane, that is, the hinge line; *i* = angle of inclination between the horizontal reference plane and the point on the commissure that is most distant from the beak; *L* = length of entire shell oriented with hinge line horizontal; *LHL* = length of the hinge line; *W* = width of a single valve from the plane of the commissure to the most inflated point on the valve.

preserved to identify and compare with known species but that yields little information on variation. A more typological systematic approach has been applied to the latter in the belief that it is more important to document the fauna at this time than to wait for the ideal sample. All indeterminate or questionably determinate forms have been mentioned and in some cases illustrated to provide the reader with a relatively complete picture of the fauna. Considering the limited exposures available to us and the intensity of our stratigraphic and paleontologic investigations in this area—basically lamina-by-lamina collecting—the faunas described here are thought to be representative of these facies in scope and preservation.

The stratigraphic occurrence of species is keyed to the localities (Fig. 1) and stratigraphic units (Fig. 3) described in Part 1 of this paper. The measured sections are described fully in Appendix 1 (see microfiche in pocket). All Oklahoma collections are in the Smithsonian Institution, U.S. National Museum, in the Department of Paleobiology (USNM numbers). U.S. Geological Survey Mesozoic locality numbers have been assigned to all collections from each section in Cimarron County. These numbers are cited with the faunals lists in section descriptions in Appendix 1. They are not repeated in the text.

SYSTEMATIC PALEONTOLOGY

Phylum COELENTERATA
Class ANTHOZOA
Order SCLERACTINIA

Ahermatypic Coral, indet.

The recrystallized calyx of a small ahermatypic coral, not retaining the interior calyx morphology and thus generically indeterminate, was noted in unit 81, section 2, upper part of the *Sciponoceras gracile* Zone, lower Bridge Creek Limestone Member, Greenhorn Formation. This type of coral is noted occasionally in dark-colored shale facies of Western Interior Cretaceous strata, but is very rare in the limestone facies, this being only the third known specimen. USNM 172864.

Phylum ANNELIDA
Class CHAETOPODA
Order TUBICOLA
Family SERPULIDAE

Genus *Serpula* Linné, 1758
"*Serpula*" sp.

Material. Encrusting tubes with worn tops and no outer ornament preserved, on *Inoceramus* shells.

Remarks. The tubes are small, 1 to 2 mm in diameter, loosely and irregularly coiled, becoming straighter terminally. They are sparsely distributed on *Inoceramus* shell fragments and without ornament are specifically indeterminate. *Serpula* is a catch-all genus used for nearly all encrusting worms found in Cretaceous strata and may include several genera by modern concepts. For this reason the name is used in quotes. Specimens are known from unit 22, section 2, lower part of Lincoln Limestone Member, Greenhorn Limestone, regional zone of *Dunveganoceras pondi*, USNM 172865.

Phylum BRACHIOPODA
Class INARTICULATA
Superfamily DISCINACEA Gray, 1840
Family DISCINIDAE Gray, 1840
Subfamily DISCINISCINAE

Genus *Discinisca* Dall, 1871
***Discinisca*? sp.**
(Pl. 6, figs. 1, 2)

Material. A single dorsal (brachial) valve with the apex missing and shell material intact. A small internal mold of a brachial valve.

Description. Small, height of largest specimen 8.2 mm, length 9.5 mm, convexity 3 mm. Conical; apex situated just below midshell, posterior bent, with flattened posterior flank, moderately curved anterior and lateral flanks; flanks moderately inclined from apex to commissure. Surface covered with fine to coarse, closely and irregularly spaced, concentric growth lines and lamellae and faint, irregularly developed concentric undulations or minor rugae (Pl. 6, fig. 1); no radial elements. Commissure flat. Shell moderately thick, lamellate, dark brown. Musculature and internal features unknown, thus generic designation questioned.

Remarks. The material is too rare and too poorly preserved for definite generic or specific identification or meaningful comparison with described Cretaceous species. The internal mold from unit 14 shows the same general form, although it has a lower apex, and is only questionably conspecific.

Occurrence. Upper part of Graneros Shale (unit 14, sec. 2), and lower part of Lincoln Member, Greenhorn Formation (unit 22, sec. 2) in slabby, hard, recrystallized, bioclastic chalky limestone or chalky calcarenite.

Specimens. Figured specimen, USNM 169309 (unit 22, sec. 2), Unfigured specimen, USNM 169333 (unit 14, sec. 2).

Phylum MOLLUSCA
Class BIVALVIA Linné (Buonanni, 1681)
Subclass PTERIOMORPHA Beurlen, 1944
Order PTERIOIDA Newell, 1965
Superfamily PTERIACEA Gray, 1847 (1820)
Family BAKEVELLIIDAE King, 1850

Genus *Phelopteria* Stephenson, 1952

For diagnosis, see Cox, 1969, p. N310, fig. C42. Also Stephenson, 1952, p. 68, Pl. 14, figs. 1–14.

***Phelopteria minuta* Kauffman and Powell, n. sp.**
(Pl. 8, figs. 4, 6)

Material. Four nearly complete, laterally compressed, small adult single valves, three right, one left, with shell partially or wholly preserved. Numerous additional fragments.

Description. Small for genus, largest observed adult attaining height of 21 mm. Subequivalve, left valve slightly more inflated than right; valves slightly biconvex with maximum inflation dorsocentral and slightly anterior, at base of umbo. Posterior, ventral flanks and auricles flattened; anterior flank slightly incurved. Moderately prosocline, average angle of inclination 42°. Beak small, blunt, slightly projecting; beak-umbo moderately prosogyrous, umbo not strongly differentiated from rest of disc. Outline subrounded (young) to subrhombic (adults) (Pl. 8, figs. 4, 6). Dorsal margin long, straight; posterior and anterior margins gently sinuous, slightly concave below auricle, slightly convex ventrally; ventral margin moderately and irregularly curved. Posterior auricle triangular, large, only slightly extended posteriorly (Pl. 8, figs. 4, 6), separated from disc by very shallow faint sulcus. Anterior auricle moderately small, triangular, prominently projecting, separated from disc by narrow, shallow, but distinct sulcus. Ornament consisting of fine to very fine, raised growth lines, very closely and irregularly spaced on juvenile shell, closely and subregularly spaced on adult portions of shell (Pl. 8, fig. 4). Shell very thin, with only the prismatic layer preserved; close set, narrow, dark brown radial color bands preserved on posterior flank of type specimen. Musculature unknown. Ligamental area poorly preserved on holotype (Pl. 8, fig. 4), seemingly a single, long, very narrow ligamental groove on small adult shells.

Remarks. This small species is widespread in the Lincoln and Hartland Members, Greenhorn Formation, and equivalent rocks of Colorado, Kansas, northeastern New Mexico, and the Oklahoma Panhandle. It has broad biostratigraphic utility within this area. It has, in the past, generally been considered a juvenile of the large Middle Turonian *Phelopteria gastrodes* (Meek), but the fact that it gets no larger than those illustrated and yet has adult hinge characters (as described by Stephenson, 1952, p. 68; Cox, 1969, p. N310) suggests that these are mature shells of a small species. Furthermore, *P. minuta* is easily distinguished from *P. gastrodes* by its greater shell inclination, lesser umbonal inflation, and much less projecting posterior auricle (compare with Stanton, 1893 [1894], Pl. 9, figs. 7-10). *Phelopteria dalli* (Stephenson, 1952, Pl. 14, figs. 4-14) has a similar disc shape but a much larger anterior auricle and somewhat more projecting posterior auricle. It is a slightly older (Early-Middle Cenomanian) species from the Woodbine Formation of Texas. *Phelopteria timberensis* Stephenson (1952, Pl. 14, figs. 1-3), also from the Woodbine Formation, is closest to *P. minuta* in disc form and development of the auricles. The disc is more erect, however, and the anterior auricle terminally rounded, with a sloping dorsal margin, compared to *P. minuta. Phelopteria salinensis* (White), the only other described middle Cretaceous species in the United States, has a higher, narrower, more inclined shell with a larger anterior auricle.

Phelopteria minuta is almost everywhere associated with calcareous and chalky shale and thin shaly limestones representing quiet-water environments below normal wave base and with dominantly very soft mud bottoms supporting a restricted benthonic fauna. Being a byssate, commonly pendant epifaunal bivalve, these probably hung on shell fragments (dominantly *Inoceramus*) or erect algal strands near the substrate surface; some may even have clung to floating rafts of seaweed, dropping into the bottom muds after death. Small pterioids were a common element of offshore Cretaceous mud communities, however (Kauffman, 1967a, p. 127-129), and are thought to have been primarily benthonic in habitat and a normal element of these assemblages.

Stratigraphic and Geographic Position. Lower middle part of the Hartland Member, Greenhorn Formation, units 11, 13, 21, 22, and 24 of section 3. Specimens questionably related to this species occur in the lower part of the Lincoln Member, unit 18, section 2, and in the upper middle part of the Hartland Member, unit 62, section 2. All specimens in buff chalky shale or thin-bedded chalky limestone.

Types. Holotype, a complete right valve (and latex cast of shell interior) with the beak worn, adult specimen, USNM 169320a, 169320b (latex cast) (unit 24, sec. 3). Unillustrated paratypes from section 3, lower part of Hartland Member, USNM 169324. Mentioned specimens, questionably this species, lower part of Lincoln Member (unit 18, sec. 2), USNM 169323; upper middle part of Hartland Member (unit 62, sec. 2), USNM 193496.

Phelopteria quadrate species aff. *P. minuta* Kauffman and Powell (Pl. 8, figs. 7, 8)

Refer to *Phelopteria minuta* Kauffman and Powell, n. sp., this paper, Pl. 8, figs. 4, 6.

Material. Two small incomplete right valves, a composite mold and specimen with shell, anterior auricles missing on both.

Description. Small, average height 14 mm; length 16.5 mm; width 1.5 mm; length of hinge line posterior to beaks, 7.5 mm; angle of inclination 54°. Slightly convex, maximum convexity dorsocentral, flanks gently sloping. Moderately prosocline, prosogyrous. Outline subcircular to subquadrate (Pl. 8, fig. 7), with straight dorsal and posterior margins, gently curved ventral and moderately rounded anterior, posteroventral margins. Posterior auricle flat, not clearly separated from disc, subtriangular, without projection of posterodorsal margin. Anterior auricle not completely known, moderately small, anteriorly projecting; with shallow notch in marginal outline below it (Pl. 8, fig. 7); beak anterocentral, blunt, rounded, very slightly projecting above dorsal margin. Ornament of very fine, crowded, subevenly spaced, slightly raised growth lines (Pl. 8, fig. 8).

Remarks. This form is clearly a new species of *Phelopteria* but is not adequately preserved or abundant enough in current collections to merit formal description and naming. It is probably descended from *P. minuta* Kauffman and Powell, n. sp., and is distinguishable from this species by its relatively longer, less elevated, more quadrate disc, by slightly coarser and more regular ornament, by having a smaller and probably less projecting anterior auricle not separated from the disc by a sulcus, and particularly in having a triangular posterior auricle that is not projecting posteriorly at its posterodorsal tip. No other described species closely resemble this form. This form is uncommon in thin-bedded shaly chalks and blocky chalks of the middle part of the Hartland Member, Greenhorn Formation, and generally occupies a stratigraphic position above *P. minuta*.

Stratigraphic and Geographic Occurrence. Units 21 and 24, section 3, and units 58 and 62, section 2, upper middle part of the Hartland Member, Greenhorn Formation, regional zone of *Dunveganoceras albertense*, Upper Cenomanian.

Figured Specimens. Composite internal mold, right valve, small adult, with good ornament and part of anterior auricle preserved, USNM 169321 (unit 24, sec. 3; Pl. 8, fig. 7); right valve, anterior auricle missing, shell preserved, small adult, USNM 169322 (unit 62, sec. 2; Pl. 8, fig 8).

Phelopteria sp. A (Pl. 8, fig. 3)

Material. A composite mold and latex cast of a small left valve with the dorsal margin, beak, and auricles incompletely preserved.

Description. Small for genus; height 16 mm, length 17.5 mm, width 2.5 mm, angle of inclination 67°. Moderately prosocline and prosogyrous; beak unknown, umbo swollen. Moderately convex (left valve), maximum inflation subcentral, anterior slope moderately inclined, posteroventral-ventral slopes slightly inclined, posterior auricle flattened, subtriangular, apparently not much extended posteriorly, separated from disc only by marked break in posterior umbonal slope. Anterior auricle triangular, flattened, markedly projecting (Pl. 8, fig. 3), well defined. Outline subquadrate; dorsal margin (hinge line) long, apparently straight, anterior margin straight, ventral and posterior margins slightly curved, ventrolateral corners moderately rounded. Surface of entire valve covered with very fine, flat, crowded growth lamellae, and moderately and irregularly spaced, prominent, asymmetrical rugae; crowding greatest dorsally (Pl. 8, fig. 3). Interior features, shell structure, unknown.

Remarks. This species is unique among described *Phelopteria* because of its coarse rugae, which easily distinguishes it from the American Cretaceous species *P. minuta*, n. sp., *P. salinensis* (White), *P. dalli* (Stephenson), *P. timberensis* Stephenson, and *P. gastrodes* (Meek). Possibly this is a pathologic specimen, or distorted by a peculiar habitat condition, but if additional and better preserved specimens are ultimately found, formal species description will be warranted.

Stratigraphic and Geographic Position. *Phelopteria* sp. A comes from unit 62, section 2, in the upper part of the Hartland Member, Greenhorn Formation, in a thin-bedded to platy, slightly resistant, buff chalky shale matrix.

Figured Specimen. USNM 169406, composite mold (a) and latex cast (b).

Family INOCERAMIDAE Giebel, 1852

General Statement Concerning Taxonomy of Inoceramid Bivalves

The ubiquitous and biostratigraphically important bivalves of the Family Inoceramidae present special problems in systematic analysis of the Oklahoma species. First, generic concepts applied to this group are used very inconsistently in the literature and are in general poorly founded. Where multiple genera are recognized within the Inoceramidae (for example, Heinz, 1932; Cox, 1969), they are largely or wholly based on external characters of the shell and do not take into account either the widespread homeomorphy in external form and ornament that has since been discovered in the family or the systematic importance of interior shell features such as the hinge plate, ligament pits, umbonal septum, musculature, and pseudodentition that are found in various inoceramid lineages. Externally very similar shells of the Inoceramidae that are now incorporated into a single genus have in many cases been shown to have entirely distinct suites of interior shell features that better define true lineage relationships. The generic- and subgeneric-level systematics of the Inoceramidae need to be completely revised, and this is currently under way by Kauffman. Until this work appears, however, the existing concepts of the *Treatise on Invertebrate Paleontology* (Cox, 1969) are used here as the basis for classification; they have been revised only where the evidence from Western Interior Inoceramidae (including those described from Oklahoma) warrants it.

Species concepts applied in the literature to Late Cretaceous Inoceramidae are similarly poor and inconsistently defined, but for a different reason. Historically, there have been, on the one hand, a great proliferation of species names among inoceramids based on typological species concepts that key on small differences in exterior shell morphology without consideration of population variation and, on the other hand, widespread lumping of species into a single generalized taxon, a procedure that assumes (without testing) considerable morphologic variation within inoceramid species. Inconsistency in use and definition of species names is great and widespread. Kauffman (1970) demonstrated that the analysis of successive, closely spaced populations within Cretaceous inoceramid lineages (including the *Inoceramus pictus* and *Mytiloides labiatus* lineages that are so common in the

Cimarron County sequence) clearly show that the morphologic variation within many species and subspecies is restricted and that the evolution of characters through time is rapid. These features can only be seen through population analysis, which has yet to be completed for most lineages. In this paper, the species concepts utilized are variable because they reflect the varying degrees to which population systematic analyses of common Cretaceous lineages are available. Refined species concepts are presented for members of the *I. pictus* and *M. labiatus* lineages because this work is virtually complete and the taxa are known; generalized, broader concepts are used for all other inoceramids not yet restudied. Kauffman is currently revising all Cretaceous Inoceramidae in the Western Interior, one lineage at a time, and this work is expected to continue for several years. The nature of this paper does not warrant full-scale revision of lineages or genera from evidence largely drawn from other parts of the world, especially where much better preserved material is available.

Genus *Inoceramus* Sowerby, 1814

For generic reference, see Cox, 1969, p. N315.

Inoceramus pictus Lineage

The dominant Late Cenomanian Inoceramidae of the Western Interior are those of the *Inoceramus pictus* Sowerby lineage. These are true *Inoceramus* (s. s.) characterized by suberect, moderately to highly inequivalve, prosocline shells with the height relatively greater than the length; moderately (right and left valves) to highly (left valve) convex shells, commonly with inflated, projecting, and incurved beaks and umbos; commonly, weak to very prominent folds and sulci posteriorly; a subtriangular posterior auricle, small to large, usually separated from the disc by an auricular sulcus; a massive hinge plate bearing large resilifers; ornament consisting of evenly to subevenly spaced, subequally developed, fine to coarse raised growth lines or flat lamellae with or without regular to irregular, weak to prominent rugae; and musculature consisting commonly of a partially pitted pallial line, comma-shaped posterior adductor, and small pedalbyssal muscle(s).

For the most part, world literature treats this group as a single, moderately variable species, applying the concept of Woods (1911, p. 279, 280), although some attempt has been made to split the lineage into distinct species and subspecies, primarily through typological systematic procedures (for example, McLearn, 1926; Warren, 1930; Sornay, 1965; Tröger, 1967). During the past few years, our systematic stratigraphic collecting and that of several colleagues in America and Western Europe have provided numerous, temporally closely spaced, large populations of the *Inoceramus pictus* lineage for analysis.

Initial observations from this study have indicated that several species and subspecies compose the *Inoceramus pictus* group in the Western Interior and Gulf Coast United States. These taxa are, for the most part, stratigraphically distinct and thus allow a refined zonation of the Middle and Upper Cenomanian (for example, Kauffman, 1970, Fig. 4). Generally parallel evolutionary trends and taxonomic sequences can be defined in the Western Interior, Gulf Coast, and Europe. Similar species occur on Madagascar, and some of the lineage is widespread throughout the Indo-Pacific region. Thus the *I. pictus* lineage is of primary biostratigraphic importance; it commonly occurs in ammonite-bearing facies, evolved at rates comparable to that of the ammonites, and had greater biogeographic and environmental spread than most ammonites at the species and subspecies level. This lineage is a useful biostratigraphic tool for global correlation of Cretaceous strata.

Primary characters useful in the systematic distinction of taxa within the lineage are shell shape, relative height, relative convexity of the valves, general inflation, development of auricles, sulci, and folds, general ornament patterns of juvenile and adult shell growth (in particular, presence or absence of rugae, the evenness of development of all concentric elements, and the similarity or differences between juvenile and adult growth), and interior features of the hinge area and muscle insertion patterns.

The early evolutionary derivation of the *I. pictus* lineage is not known precisely although it is generally thought that this lineage arose from a branch of the Albian "*Inoceramus*" *comancheanus* Cragin lineage, possibly through variants of "*I.*" *anglicus* Woods. The lineage diversified into two main branches (*I. pictus, I. tenuistriatus*) in Cenomanian time, and both branches terminated in America at the end of the Cenomanian insofar as is known; *Inoceramus cuvieri* Lamarck, the genotype from the Lower Turonian, is a probable offshoot of the lineage.

Inoceramus pictus pictus Sowerby (Pl. 1, fig. 1)

Refer to *Inoceramus pictus* Sowerby, 1829, p. 215, Pl. dciv, fig. 1. For partial synonymy, see Woods, 1911 (part), p. 279, 280, Fig. 36, non Pl. 49, figs. 5, 6.

Diagnosis. Moderate size for genus; moderately biconvex with left valve significantly more inflated than right, especially in beak-umbo area; maximum inflation dorsocentral below umbo. Suberect, slightly to moderately prosocline; height markedly greater than length. Hinge line short, straight; anterior margin straight to slightly concave; posterior margin straight to slightly convex ventroposteriorly, with slight concave re-entrant at base of posterior auricle. Auricle triangular, flattened, of moderate size, slightly projecting in posterior direction along hinge line, separated from disc by shallow auricular sulcus. Ventral margin moderately and asymmetrically rounded, with slight posteroventral flattening in margin and growth line trace, in some specimens marking flat area or very shallow sulcus along anterior flank of low umbonal fold. Beak, umbo suberect to somewhat prosogyrous, moderately to strongly inflated on right and left valves, respectively, projecting dorsally, with left beak more prominent; beaks strongly incurved over hinge line, pointed; apical angle acute. Ornament consisting of medium size, equally developed, closely and subevenly spaced, raised rounded and crowded growth lines becoming finer on umbo and weaker on gerontic shell areas. Moderately spaced, medium-sized rugae and concentric constrictions are subregularly distributed over shell, and become more crowded ventrally on adult shells. Hinge area moderately thickened, containing multiple resilifers shaped like vertically elongate rectangles; closely spaced with narrow, broadly rounded interspaces. Pallial line entire, weakly developed. Adductor muscle impression weak, laterally elongated parallel to trend of the pallial line (boat-shaped); no pedalbyssal muscles yet observed, although small dorsal ones present in some other members of lineage. Umbonal septum small, curved, moderately thickened, not strongly notched. Shell moderately thin except in ligamental plate.

Remarks. *Inoceramus pictus pictus* Sowerby has not yet been found in Oklahoma although it does occur in Upper Cenomanian strata elsewhere in the Western Interior and in the Eagle Ford Formation of the Texas Gulf Coast. It is described and illustrated here for comparison with related taxa which have been formerly ascribed to this species in North America and because it forms an important biostratigraphic index in the general scheme of inoceramid zonation.

As can be observed on Plates 1 through 4, *I. pictus pictus* can be easily distinguished from related species and subspecies occurring in Oklahoma and elsewhere in the Western Interior, as follows: It is more inflated, and has coarser growth lines, more projecting beaks and umbos, a steeper anterior margin that has a concave trace, and a more concave posterior margin below the auricle than the subspecies *I. pictus gracilistriatus*, n. subsp.; *I. pictus pictus* has finer and more even growth lines (especially on the adult ventral portion of the shell), a more prominent beak and umbo, and more regular development of rugae than *I. flavus pictoides* Sornay (Pl. 1, fig. 4). *I. prefragilis prefragilis* Stephenson and the new subspecies *I. pictus stephensoni* (Pl. 2, figs. 8, 10) lack the regular rugae of *I. pictus pictus*, are less convex, and have coarser raised growth lines. The subspecies *I. pictus prefragilis* further possesses a posteroventrally trending shallow sulcus. *I. ginterensis* (Pl. 4) has much more irregular ornament, is broader and much less convex, and has more depressed beaks and umbos. *Inoceramus* aff. *I. tenuis* and *I. tenuistriatus*? from Oklahoma, which seem to belong to a distinct lineage but have been formerly identified with *I. pictus*, are both much more quadrate, less convex shells with finer concentric ornamentation and depressed beaks and umbos.

Stratigraphic Position. *Inoceramus pictus pictus* is, in the restricted sense used here, limited to Late Middle and Late Cenomanian rocks of North America and Western Europe.

Type. Holotype, British Museum no. 43272, a bivalved specimen with the beaks broken off and part of the posterior auricle missing. The specimen illustrated in Plate 1, figure 1 is a plastoholotype at the U.S. National Museum, USNM 169393.

Inoceramus pictus gracilistriatus Kauffman and Powell, n. subsp. (Pl. 1, fig. 3)

Inoceramus pictus neocaledonicus Jeannet. Pergament, 1966, p. 55–57 (part), Pl. 30, figs. 2, 3, 4(?); Pl. 29, figs. 3(?), 4(?). non *Inoceramus neocaledonicus* Jeannet, 1922, p. 251–253, Fig. 5 on p. 252.

Material. A single, nearly complete, internal mold of a left valve, beak missing, ventral margin broken.

Description. Moderate size for genus; height 54 mm, length 38 mm, width 8 mm. Prosocline, angle of inclination 72°; slightly prosogyrous, apical angle 87°. Moderately convex (left valve), maximum inflation subcentral, just dorsal to midshell; anterior flank moderately to steeply inclined; ventral and posterior flanks gently sloping. Posterior auricle very slightly sloping, subtriangular with rounded apex (Pl. 1, fig. 3), moderate size, differentiated from disc by moderate break in postumbonal slope. No sulci; umbonal fold low and indistinct. Beak presumably blunt, beak-umbo only slightly projecting above hinge; umbo somewhat depressed with relatively low convexity. Shell suberect, outline elongate-ovate, height one-third greater than length; anterior and posterior margins slightly curved convexly outward; posterodorsal margin of moderately short length (20 mm), straight; ventral margin moderately and unevenly curved (Pl. 1, fig. 3).

Ornament consisting of weak to moderately coarse rugae, sparse and irregularly developed dorsally, moderately and subevenly spaced ventrally; first ruga at base of umbo (17-mm axial length, juvenile shell; Pl. 1, fig. 3); one ruga in first 25-mm axial length, three rugae in second 25-mm axial length. Internal mold and presumably shell covered with fine to very fine, subequally developed, closely and subregularly spaced, raised rounded growth lines (Pl. 1, fig. 3), about 33 in first 25-mm axial length, 25 in second 25-mm axial length. Growth lines becoming weaker ventrally and posteriorly on adult shell.

Prismatic layer of ligamental plate somewhat thickened, layer otherwise very thin. Nacre not preserved; resilifers unknown. No pedalbyssal or adductor muscle insertion areas preserved. Pallial line very thin, faint, submarginal, presumably entire, incompletely known (Fig. 2). Posterior adductor track moderately broad, straight, situated on umbonal fold. Faint radial tracks of mantle retractor muscles cover dorsal part of shell.

Remarks. *Inoceramus pictus gracilistriatus*, n. subsp., is similar in most ways to *I. pictus pictus* Sowerby (Pl. 1, fig. 1) but easily distinguished from the typical form by its finer and more irregular growth lines, somewhat fewer, more widely spaced rugae, lower convexity, and slightly less pronounced posterior auricle. The subspecies is identical to forms identified by Pergament (1966, especially Pl. 30, fig. 4) as *I. pictus neocaledonicus* Jeannet but is more erect and much more finely ribbed than the type of that species (Jeannet, 1922, p. 251–253, fig. 5). The younger *Inoceramus tenuistriatus*? Nagao and Matsumoto from Oklahoma has closer, more regular, less pronounced rugae, a shorter, more quadrate shell, finer, fainter growth lines, and a more inflated, more prosogyrous beak-umbo (Pl. 3, figs. 7, 9, 10). *Inoceramus pictus gracilistriatus* is a rare species in the lower part of the Hartland Member, Greenhorn Formation, of Oklahoma but is common at this level in Colorado. No morphologic gradations exist between this subspecies, *I. flavus pictoides* (much coarser ribbed), and *I. ginterensis* (more irregularly and more coarsely ribbed, flatter), with which it co-occurs, and it is best at present to separate them taxonomically, as did Pergament (1966).

Stratigraphic and Geographic Position. Unit 14, section 3; lower part of Hartland Member, Greenhorn Formation, in moderately resistant, light-colored, chalky limestone.

Type. Holotype, internal or composite mold of a nearly complete, adult left valve, USNM 169394.

Inoceramus flavus pictoides Sornay
(Pl. 1, fig. 4; Pl. 2, fig. 4)

Inoceramus flavus var. *pictoides*, Sornay, 1965, p. 6, 7.

Inoceramus pictus Sowerby. Woods, 1911, p. 279 (part). Donovan, 1954, p. 17, Pl. 3, figs. 5, 6(?). Other authors in part.

Cricoceramus (*Cricoceramus*) *pictus* Sowerby. Heinz, 1933 (1934), p. 245, 246, Pl. 16, fig. 3, non fig. 4.

?*Inoceramus pedalionoides*, Nagao and Matsumoto, 1939, Pl. 26(4), figs. 8a–c, 9(?).

Material. Internal or composite molds of adult right and left valves, plus large fragments.

Description. Table 1 gives summary of measurements. Moderate size for genus, maximum recorded height about 75 mm (right valve). Moderately prosocline, average angle of inclination 71°. Moderately, subequally biconvex, left valve slightly larger, especially in beak projection (Pl. 2, fig. 4), and probably more inflated. Maximum inflation dorsocentral just below base of umbo; anterior flank moderately inclined, ventral and posterior flanks gently sloping. Posterior auricle flattened, triangular, of moderate size, separated from disc by marked break in postumbonal slope (Pl. 1, fig. 4). Slightly prosogyrous, apical angle averages 90°; beak narrowly rounded,

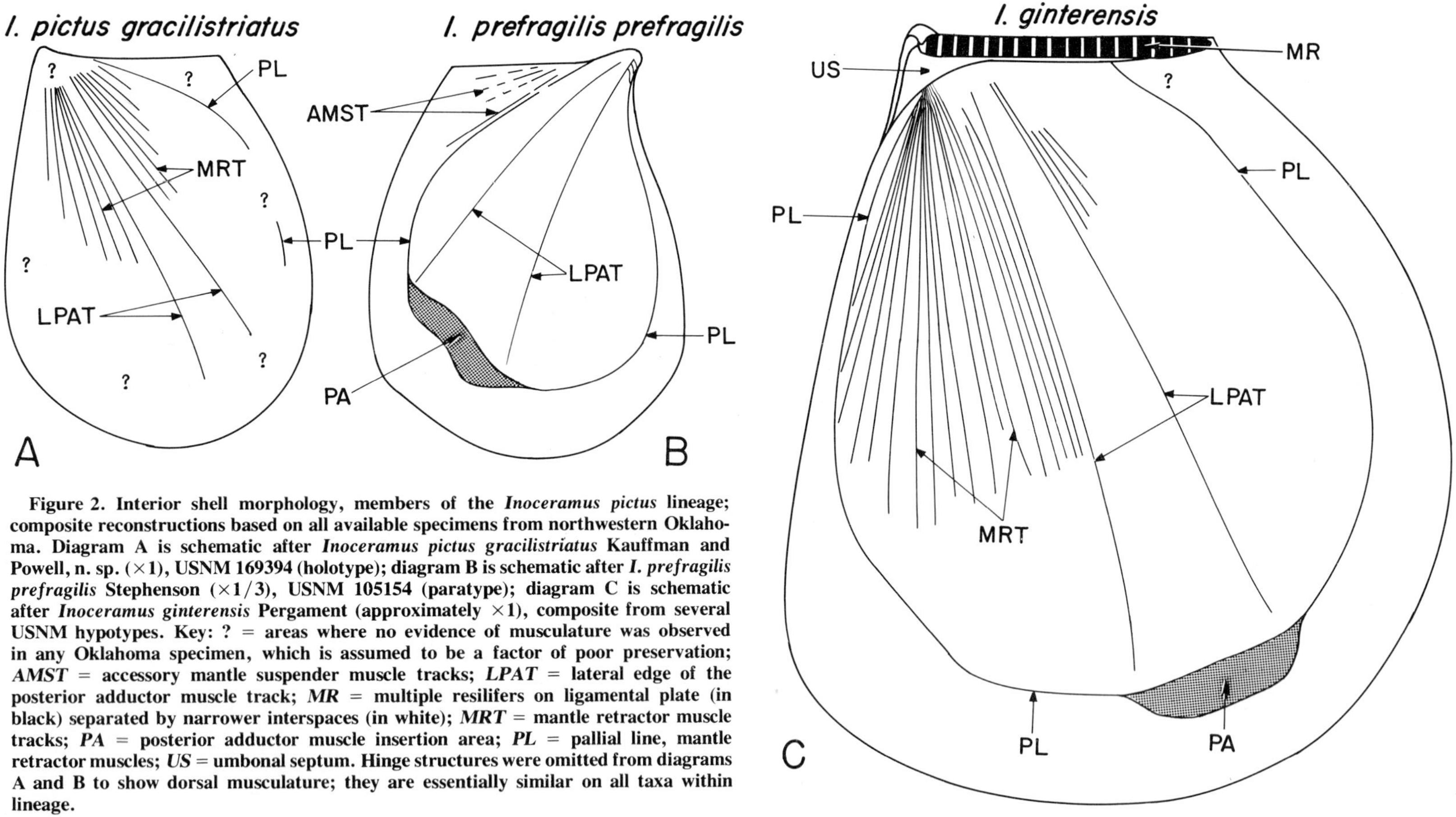

Figure 2. Interior shell morphology, members of the *Inoceramus pictus* lineage; composite reconstructions based on all available specimens from northwestern Oklahoma. Diagram A is schematic after *Inoceramus pictus gracilistriatus* Kauffman and Powell, n. sp. (×1), USNM 169394 (holotype); diagram B is schematic after *I. prefragilis prefragilis* Stephenson (×1/3), USNM 105154 (paratype); diagram C is schematic after *Inoceramus ginterensis* Pergament (approximately ×1), composite from several USNM hypotypes. Key: ? = areas where no evidence of musculature was observed in any Oklahoma specimen, which is assumed to be a factor of poor preservation; *AMST* = accessory mantle suspender muscle tracks; *LPAT* = lateral edge of the posterior adductor muscle track; *MR* = multiple resilifers on ligamental plate (in black) separated by narrower interspaces (in white); *MRT* = mantle retractor muscle tracks; *PA* = posterior adductor muscle insertion area; *PL* = pallial line, mantle retractor muscles; *US* = umbonal septum. Hinge structures were omitted from diagrams A and B to show dorsal musculature; they are essentially similar on all taxa within lineage.

umbo moderately inflated, not clearly separated from main part of shell. Outline elongate-ovate (Pl. 1, fig. 4), shell erect, with height markedly greater than length. Dorsal margin posterior to beak straight, short; anterior and posterior margins gently curved convexly; ventral margin moderately and subevenly rounded; posteroventral margin slightly flattened, defined by growth line trace (Pl. 2, fig. 4).

Ornamentation similar in adult and juvenile shell, consisting of closely and subevenly spaced, subequally developed, coarse raised growth lines, becoming weaker and more distantly spaced near ventral, posterior margins of adult shells (Pl. 2, fig. 4). Sparse, widely and irregularly spaced, weak to moderate rugae (Pl. 1, fig. 4) first appear near midshell, becoming weaker but more crowded ventrally.

Ligamental plate moderately thickened, resilifers unknown. No pedalbyssal musculature observed. Pallial line entire insofar as it is preserved (Fig. 3), faint, very thin, slightly sinuous. Posterior adductor insertion area (Fig. 3) known only from ventral margin; laterally elongate, probably subcrescentic in outline; situated posteroventrally near commissure. Only prismatic layer preserved, very thin except near ligamental plate.

Remarks. *Inoceramus flavus pictoides* partially co-occurs with *I. ginterensis* Pergament and closely resembles it in general form, juvenile ornament, and distribution of rugae. *Inoceramus flavus pictoides* differs from *I. ginterensis* primarily in having coarse, subequally developed, raised growth lines over the entire shell, in having fewer and generally weaker rugae, and in being somewhat higher relative to length. Whereas these differences are not of great significance in some highly variable species of *Inoceramus*, no morphologic gradients exist between these two in the present collections, and it seems best to consider them as distinct taxa until such gradation can be established from larger collections.

Among the many described taxa of the *Inoceramus pictus* group, this form most closely resembles *I. pedalionoides* Nagao and Matsumoto, from Upper Cenomanian and Turonian strata of Japan. The two are similar in form and the general nature of the ornament; they differ mainly in the greater development of the posterior auricle and coarser adult rugae and in the greater number of rugae on the Japanese species. *I. pedalionoides* also has somewhat finer growth lines and a concave anterior margin, as typically developed.

Inoceramus pictus, with which this form has been previously identified in the Western Interior, has more numerous and regular rugae, finer growth lines, a less evenly curved growth line trace, and a slightly concave posterior margin that

Table 1. Summary of Measurements for *Inoceramus flavus pictoides* Sornay

Character*	USNM 169395 right valve	USNM 169400 left valve	Average
H	75 mm	61 mm	68.0 mm
L	53 mm	40 mm	46.5 mm
W	8 mm	5 mm	7.5 mm
LHL	22 mm	20 mm	21.0 mm
i	73°	69°	71°
a	93°	87°	90°
R/AL_1	0	1	0.5
R/AL_2	2	1	1.5
R/AL_3	2	. .	. .
GL/AL_1	31	27	29
GL/AL_2	15	17	16
GL/AL_3	13	. .	. .

*Character measurements explained in Figure 1. R = number of rugae. GL = number of growth lines. AL_1 = first 25 mm of axial length. AL_2 = second 25 mm of axial length. AL_3 = third 25 mm of axial length.

suggests development of a larger, posteriorly extended auricle (compare Pl. 1, fig. 4, with Pl. 1, fig. 1—the type of *I. pictus*). *I. flavus pictoides* is probably derived from *I. prefragilis stephensoni*, n. subsp., which it resembles closely. It is distinct from the older form in having well-defined rugae, a less rounded growth line trace, more irregular growth lines ventrally and laterally, and a convexly curved anterior margin. Compared with *I. flavus pictoides*, *I. pictus etheridgei* (Etheridge, Jr.) has a much more prominent prosogyrous beak-umbo and more irregular ornament, *I. pictus vardonensis* Sornay has much coarser growth lines, regular rugae, and a relatively higher shell, and *I. prefragilis* Stephenson has more regular ornament over the entire shell, generally lacks prominent sulci, and has a faint posteroventrally directed sulcus (see Pl. 1, figs. 5, 6).

Stratigraphic and Geographic Position. Regional zone of *Dunveganoceras albertense*. Lower and middle parts of Hartland Member, Greenhorn Formation; unit 50, section 2; units 2 and 3, section 3; Cimarron County, Oklahoma. A poorly preserved, possibly related specimen occurs in unit 27, section 2, in the middle part of the Lincoln Member.

Hypotypes. Adult right valve, internal mold, USNM 169395 (unit 3, sec. 3). Laterally crushed adult left valve, auricle incomplete, composite internal mold, USNM 169400 (unit 3, sec. 3). Fragment of dorsal half of left valve, USNM 169408 (unit 50, sec. 2).

Inoceramus prefragilis prefragilis Stephenson (Pl. 1, figs. 5, 6; Pl. 2, figs. 7, 9)

Inoceramus prefragilis Stephenson, 1952, p. 64, 65, Pl. 12, figs. 10–12; Pl. 13, figs. 1, 2.

Diagnosis. Stephenson (1952, p. 64, 65) provided a full description and illustrated the range of variation, which we have supplemented by illustrations of additional paratypes in this paper. Medium to large, attaining height of 147 mm. Moderately prosocline and prosogyrous. Shell suberect, length one-third to one-quarter less than height. Moderately inequivalve; moderately biconvex, left valve somewhat more inflated and with more projecting beak-umbo than right. Beak narrowly rounded; umbo inflated, moderately curved inward and anteriorly. Outline erect (Pl. 1, fig. 5), elongate-ovate; dorsal margin straight, of moderate length; anterior margin consistently steep, with concave trace; lunule faint; posterior margin slightly and convexly curved, flattened at auricle. No anterior auricle; posterior auricle triangular, well defined, separated from disc by shallow sulcus and (or) distinct break in postumbonal slope. Umbonal fold low, asymmetrically rounded; a shallow but distinct posteroventrally trending sulcus lies parallel and behind umbonal fold (Pl. 2, fig. 9) and becomes more prominent in adult shells. Ornament of coarse, subequally developed, evenly to subevenly spaced, raised rounded growth lines and very faint to moderately prominent, irregularly developed rugae, sparse on young adults, crowded at older growth stages; juvenile ornament coarse and similar to that of adult. Hinge characters illustrated by Stephenson (1952, Pl. 13, fig. 2); musculature shown in Figure 2.

Remarks. The holotype and young specimens illustrated by Stephenson (1952, Pl. 12, figs. 10–12, Pl. 13, figs. 1, 2) do not adequately represent the species or show its typical form. More typical specimens from Stephenson's suite of paratypes are illustrated here (Pl. 1, figs. 5, 6; Pl. 2, fig. 9). The shallow but distinct umbonal and auricular sulci are characteristic of the species population and serve to distinguish it from most others in the lineage. *I. prefragilis prefragilis* is the oldest known American representative of the species group. Compared with

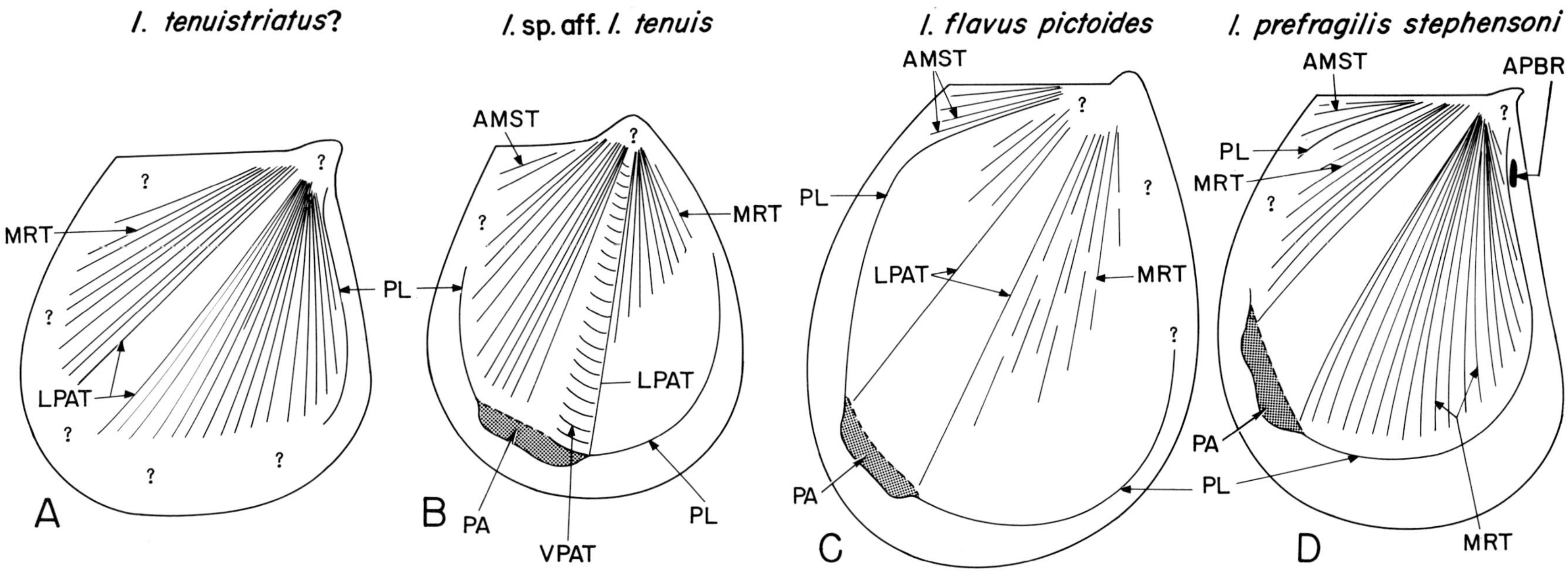

Figure 3. Interior shell morphology, members of the *Inoceramus pictus* and *I. tenuistriatus* lineages in northwestern Oklahoma. Unless otherwise specified, drawings are schematic composites from all available specimens with hinge (ligamental) plates and umbonal septum removed to show dorsal musculature (Fig. 2C is typical of these for both lineages). Diagram A is schematic after *Inoceramus tenuistriatus*? Nagao and Matsumoto (×1), musculature composite of USNM 169399, 169402, 169403; diagram B is schematic after *Inoceramus* sp. aff. *I. tenuis* Mantell (×1 1/4), primarily from USNM 169405; diagram C is schematic after *Inoceramus flavus pictoides* Sornay (×1) after USNM 169395, from hypotypes USNM 169395, 169400; diagram D is schematic after *Inoceramus prefragilis stephensoni* Kauffman and Powell, n. subsp. (×1), USNM 169325 (holotype); musculature composite from holotype and paratypes USNM 169326b, 169327. Key: ? = areas where no evidence of musculature was preserved on any specimen and thus does not indicate nonoccurrence of muscles, *AMST* = accessory mantle suspender muscle tracks; *APBR* = possible anterior pedalbyssal retractor muscle insertion area (on one specimen only); *LPAT* = lateral edge of posterior adductor muscle track; *MRT* = track of mantle retractor muscles that make up pallial line; *PA* = posterior adductor muscle insertion area; *PL* = pallial line; *VPAT* = trace of successive ventral margins of the posterior adductor muscle track.

I. pictus Sowerby (Pl. 1, fig. 1), *I. prefragilis prefragilis* has coarser growth lines, weaker and fewer rugae, a smaller posterior auricle that does not project posteriorly at its outer edge and loses its ornament earlier in adult growth (see Stephenson, 1952, Pl. 13, fig. 1). Other forms are easily distinguished from *I. prefragilis prefragilis*. The new subspecies *I. prefragilis stephensoni* lacks sulci, is relatively longer, and has a rounded growth line trace. *I. flavus pictoides* lacks sulci and has somewhat finer growth lines, more numerous and coarser rugae (especially in early adult growth), and a convexly curved anterior margin. *I. pictus gracilistriatus*, n. subsp., has much finer growth lines and regular, more numerous rugae.

Inoceramus prefragilis prefragilis Stephenson is not known yet from Oklahoma, although Middle Cenomanian rocks (Graneros Shale) that are equivalent in age to its Texas occurrence are found here. *I. prefragilis stephensoni* is descended from the Texas form and occurs in slightly younger Cenomanian strata. The species is common in the Lincoln and Hartland Members, Greenhorn Formation, of Kansas.

Stratigraphic and Geographic Occurrence. Middle Cenomanian zone of *Calycoceras (Conlinoceras) tarrantense* and zone of *Metoicoceras swallovi, M. latoventer, M. crassicostae*, Lewisville and Templeton Members, respectively, of the Woodbine Formation at various localities in Tarrant, Denton, Cook, Grayson, Fannin, Lamar, and Red River Counties, Texas, described by Stephenson (1952). The species has also been found in the lowest part of the Eagle Ford Formation in Texas.

Types. Holotype, USNM 105151; figured paratypes, USNM 105152a, 105152b, 105154 (Stephenson, 1952, p. 65), Lewisville Member, Woodbine Formation, near a small unnamed branch 1.70 km (1.05 mi) east and 0.32 km (0.2 mi) south of Penland, Terrace Station, 0.5 km (0.3 mi) southeast of Dugans Chapel, Grayson County, Texas. Zone of *Calycoceras (Conlinoceras) tarrantense*. Newly illustrated paratypes from the same locality: a large right valve with unusually weak rugae and faint sulci, USNM 169314 (Pl. 1, fig. 5); typical small adult left valves, USNM 169315 (Pl. 1, fig. 6), USNM 169317 (Pl. 2, fig. 7); an unusually coarsely rugate adult left valve with prominent sulci, USNM 169316 (Pl. 2, fig. 9).

Inoceramus prefragilis stephensoni Kauffman and Powell, n. subsp. (Pl. 2, figs. 1, 3, 6, 8, 10)

Material. Six right and five left adult valves, 50 to 90 percent complete, internal molds, some with shell adhering.

Description. Summary of measurements in Table 2. Moderate size for genus, attaining height of 67 mm. Moderately inequivalve, left valve more inflated and slightly larger than right, difference most pronounced in umbonal area; left beak slightly overhangs and projects beyond right beak. Moderately biconvex, greatest inflation dorsal and anterior to midpoint of disc at base of umbo; anterior flank steep to vertical, ventral and posterior flanks slightly inclined to flattened (Pl. 2, fig. 10). Shallow, indistinct lunule developed below beak in some specimens, bounded by a low rounded ridge (Pl. 2, fig. 3). Moderately prosocline, average angle of inclination 72°. Prosogyrous, with curved umbonal axis. Beaks blunt, narrowly rounded, terminal anteriorly, only slightly projecting dorsally and anteriorly on right valves, moderately projecting on left valves. Posterior auricle subtriangular, not clearly separated from disc by sulcus, but by gentle change in posterior umbonal slope. Valve outline subovate to subrectangular, with long straight posterodorsal hinge margin, straight to slightly concave anterior margin, slightly curved posterior margin, and moderately convex ventral edge (Pl. 2, fig. 10).

Ornament consisting of coarse, rounded, subequally developed, closely and subevenly spaced, raised growth lines over whole shell, becoming finer and somewhat

Table 2. Summary of Measurements for *Inoceramus prefragilis stephensoni*, n. subsp.

Character*	Average	USNM 169395 holotype	USNM 169326b paratype	USNM 169327 paratype	USNM 169328a paratype†	USNM 169328b paratype†
Valve		Right	Right	Right	Right	Left
H	53.2 mm	66 mm	57 mm	54 mm	42 mm	47 mm
L	38.6 mm	50 mm	41 mm	30 mm	35 mm	37 mm
W	8.8 mm	12 mm	8 mm	9 mm	6 mm	9 mm
LHL	20.2 mm	30 mm	18 mm	16 mm	. .	17 mm
i	72°	65°	72°	76°	73°	74°
GL/AL_1	22	21	25	26	20	18
GL/AL_2	15	11	17	17	17	13
AL/R_1	34 mm §	0	23 mm	45 mm	0	0
AL/R_2	48.5 mm §	0	45 mm	52 mm	0	0

*Character measurements explained in Figure 1. GL = number of growth lines. AL_1 = first 25 mm of axial length. AL_2 = second 25 mm of axial length. AL = axial length. R_1 = first ruga. R_2 = second ruga.

† Unfigured.

§ Where present.

less regular on umbo. Growth lines with rounded, even, concentric curvature (Pl. 2, fig. 10), very slightly flattened ventroposteriorly on one specimen (Pl. 2, fig. 6). Very faint, sparse, shallow, incomplete rugae occur on about one-half of specimens, mainly near margins of larger shells and thus probably a gerontic feature (Pl. 2, fig. 6). No umbonal or auricular sulcus, as found in *I. prefragilis prefragilis* Stephenson. Shell thin, somewhat thickening near hinge, where prismatic layer (only one preserved) reaches 1 mm in thickness. Hinge characters unknown on Oklahoma specimens, presumably like those of subspecies *I. prefragilis prefragilis* (Stephenson, 1952, Pl. 13, fig. 2). Muscle insertion areas poorly preserved. Pallial line faint, entire, slightly wavy ventrally. Faint, closely spaced, mantle retractor muscle tracks extend from umbo to pallial line over all of shell interior except posterior adductor muscle track, which is moderately narrow, situated on the umbonal slope, and terminated in a laterally elongate, bipartite(?), posterior adductor insertion area, poorly defined, dorsal margin of area unknown. No obvious pedalbyssal musculature on preserved specimens with possible exception of small elongate scar adjacent to ligamental area dorsoanteriorly (Fig. 3).

Remarks. This subspecies is derived evolutionarily from *I. prefragilis prefragilis* Stephenson (Pl. 1, figs. 5, 6; Pl. 2, figs. 7, 9), a Middle and early Late Cenomanian species of the Woodbine Formation in Texas. In most respects the two forms are identical. Subspecies *I. prefragilis prefragilis*, however, is slightly more convex, with greater relative height, less prosogyrous but more projecting beaks and umbos, somewhat coarser ornament on the average (34 growth lines in 50 mm compared to 39), and possesses, in most specimens, a shallow umbonal sulcus, marked by a distinct flattening of the growth line trace, between the posterior umbo and posteroventral margin (Pl. 2, fig. 9). These differences are consistent in collections from outside Oklahoma, from Texas to Colorado, and easily recognized in good material, justifying description of a new subspecies. Very large adult specimens (to 30-cm height) from Colorado probably belonging to subspecies *I. prefragilis stephensoni* maintain regular concentric ornament throughout growth, while large adults of subspecies *I. prefragilis prefragilis* (Stephenson, 1952, Pl. 13, fig. 1) lose regular growth lines in old growth stages and develop instead numerous, faint, irregular rugae. *Inoceramus prefragilis stephensoni* marks a distinct, early Late Cenomanian zone in the Western Interior and Gulf Coast, between subspecies *I. prefragilis prefragilis* and younger members of the *I. pictus* group.

The subspecies is named in honor of the late L. W. Stephenson of the U.S. Geological Survey, who first studied this lineage in the United States.

Geographic and Stratigraphic Position. Lower part of Lincoln Member, Greenhorn Formation, regional zone of *Dunveganoceras pondi,* units 18, 20, 22, 24, 26, and 32 (poor specimens), section 2; in slabby to blocky calcarenites and calcarenitic limestones.

Types. Holotype, internal mold of adult right valve with partial shell adhering, ventrolateral margin broken, USNM 169325 (unit 22, sec. 2). Paratype, right valve, internal mold, hinge area missing, USNM 169327 (unit 22, sec. 2). Paratype, external mold and latex cast, right valve, auricle broken, USNM 169326a, 169326b (latex) (unit 22, sec. 2). Paratype, internal mold of left umbonal area, USNM 169396; mold and latex cast of right valve umbo, USNM 169397a, 169397b. Unillustrated paratypes (all sec. 2): USNM 169328a, 169328b (unit 22); USNM 169329 (unit 18), USNM 169330 (unit 20), USNM 169331 (unit 26), USNM 169332 (unit 32).

Inoceramus ginterensis Pergament
(Pl. 1, fig. 2; Pl. 3, figs. 1–4, 6; Pl. 4, figs. 1, 3–6; Pl. 5, figs. 2, 3, 5; Pl. 7, fig. 4)

Inoceramus ginterensis Pergament, 1966, p. 50–52, Pl. 25, fig. 5; Pl. 26, figs. 1, 2; Pl. 27, figs. 1, 2; Pl. 28, figs. 1, 2; Pl. 29, fig. 1.

Inoceramus flavus Sornay, 1965, p. 4 (part), Pl. A, fig. 1; non Pl. A, fig. 2; non Fig. 2.

Material. Thirty well-preserved, mostly complete, internal and composite molds of single right and left valves, mostly adults, some with portions of prismatic shell adhering. Numerous large fragments and poorly preserved internal molds of adult valves. All growth stages well represented.

Description. Summary of measurements given in Table 3. Moderate size for genus, attaining nearly 12-cm shell height on whole specimens, with larger fragments indicating nearly twice this size. Moderately prosocline to suberect, average angle of inclination 66.1°. Inequivalve, left valve slightly more inflated, slightly higher than right; slightly biconvex, low for lineage, maximum inflation in dorsoposterior quadrant near base of umbo. Anterior margin moderately steep, ventral and posterior flanks flattened, slightly inclined (Pl. 4, fig. 4). Posterior auricle moderate in size, subtriangular, poorly defined, separated from disc only by subtle break in posterior umbonal slope; no auricular sulcus. Outline elongate-ovate to subrectangular (Pl. 4, figs. 3–6; Pl. 3, figs. 1–4, 6), with moderately long, straight hinge line posterodorsally, broadly curved (convexly) posterior margin, moderately and unevenly curved ventral margin, and straight to very slightly curved (concave or convex) anterior margin. Beaks anterior, terminal, blunt to narrowly rounded, very slightly projecting above hinge line; apical angle somewhat obtuse.

Juvenile ornament to 10- to 20-mm shell height consisting of closely and subevenly spaced, raised, moderately coarse, subequal growth lines (Pl. 3, figs. 2, 4). Adult ornament irregular, consisting of coarse, moderately and irregularly (Pl. 4, figs. 4, 5) to subregularly (Pl. 3, fig. 6) spaced rugae, between and on which are faint to moderately developed, unequal, moderately to closely and irregularly spaced, raised growth lines (Pl. 4, figs. 4–6 are typical). Trace of ornament typically rounded, in some cases slightly flattened along low umbonal fold (Pl. 4, fig. 5). Some large specimens with nearly smooth posterior flanks (Pl. 5, fig. 3), coarser ornament centrally and anteriorly.

Ligamental plate thickened with numerous, moderately deep, concave, rectangular resilifers (Pl. 4, fig. 1) separated by much narrower, broadly rounded to flat-topped, ligamental platforms; seven resilifers in 10-mm length along hinge line. Small, triangular umbonal septum developed below resilifers, as a ventral projection of anterior one-half of the ligamental plate (Pl. 4, fig. 1). Very low, internal rib incompletely developed dorsoposteriorly along junction of auricle and disc. Pallial line entire, narrow; trace shown in Figure 2. Posterior adductor muscle track faintly defined by low bounding ridges (grooves on internal mold; Pl. 4, fig. 5), moderately broad, trending along umbonal fold from umbo to ventroposterior margin. Adductor muscle insertion area laterally elongate (Fig. 2), shallow pan-shaped, with flat base. No pedalbyssal musculature observed. Nacre not preserved. Prismatic layer very thin (0.5 to 0.75 mm) over most of shell, moderately thickened for genus at ligamental plate (to 3 mm on large specimens).

Ontogenetic variation in ornament and shape as follows: abrupt ornament change from regular juvenile ornament—equally developed raised growth lines (Pl. 3, fig. 2)—to adult ornament with fainter, more irregular growth lines and irregular to subregular rugae (Pl. 4, figs. 4-6). Adult growth lines becoming fainter, and rugae generally more crowded with age (Pl. 4, figs. 4, 5 are typical). Shape change from subquadrate to ovate young valves (Pl. 1, fig. 2; Pl. 3, fig. 1) to higher, more erect adults (Pl. 4, figs. 4-6) as ventral growth exceeds lateral growth.

Adult variation primarily reflected in (1) ratios of height to length, from tall relatively narrow shells (Pl. 4, fig. 6) to broader forms (Pl. 4, fig. 4) with height only slightly greater than length, and (2) in ornament, particularly the number and coarseness of rugae (compare Pl. 4, fig. 4, with Pl. 5, fig. 3) and their relative development posteriorly (compare Pl. 5, fig. 3, with Pl. 4, figs. 4-6).

Remarks. *Inoceramus ginterensis* Pergament represents a cosmopolitan lineage, occurring commonly in the Cenomanian rocks of the Atlantic and Indo-Pacific areas. Slight differences in local populations have led to the formulation of numerous species whose validity has not been tested against world-wide population analysis. Of these, *I. ginterensis* is identical with most Oklahoma forms, whereas other "species" show slight and possibly consistent differences. These forms have been referred previously to either *Inoceramus pictus* Sowerby or *I. prefragilis* Stephenson and clearly differ from both in the irregularity of the growth lines and rugae and

Table 3. Summary of Measurements for *Inoceramus ginterensis* Pergament

Character*	Valve†	No. of specimens	Range	Average
H	R	15	9-103 mm	62.1 mm
H	L	10	42-87 mm	63.8 mm
L	R	14	8-80 mm	46.7 mm
L	L	10	32-67 mm	51.0 mm
W	R	14	2-14 mm	7.4 mm
W	L	9	5-8 mm	6.9 mm
LHL	R and L	19	5-41 mm	23.8 mm
i	R and L	24	54-82°	66.1°
a	R and L	23	65-93°	84.0°
R/AL_1	R and L	24	0-4	1.5
R/AL_2	R and L	20	1-3	2.2
R/AL_3	R and L	11	1-3	2.0
R/AL_4	R and L	6	0-3	1.8
GL/AL_1	R and L	18	13-26	19.6
GL/AL_2	R and L	20	8-15	12.5
GL/AL_3	R and L	11	8-13	10.4
GL/AL_4	R and L	5	5-12	9.2

*Character measurements explained in Figure 1. R = number of rugae. GL = number of coarse raised growth lines. AL_1 = first 25 mm of axial length. AL_2 = second 25 mm of axial length. AL_3 = third 25 mm of axial length. AL_4 = fourth 25 mm of axial length.

†Right (R); left (L).

in having less convex, less inequivalve shells.

Closely related, similar forms are *I. flavus* Sornay, from Madagascar, *I. pictus etheridgei* (Etheridge, Jr.) from eastern (including Pacific) and western Europe, and *I. beringensis* Pergament from the Pacific coast of the USSR. Sornay (1965, Pl. A, fig. 1) included in his concept of *I. flavus* a form probably conspecific with *I. ginterensis*, showing subequal growth lamellae on the shell exterior (not seen on Oklahoma specimens, lacking shell, but clearly developed on *I. ginterensis* from the USSR) but otherwise identical characteristics to Oklahoma forms. This is not the holotype of *I. flavus* though, which is distinct from *I. ginterensis* in being more prosocline, and much more prosogyrous and in having closely spaced, subequally developed, raised growth lines and small rugae. *I. pictus etheridgei* has a much more projecting beak and umbo, coarser and more equally developed, raised growth lines, and is more prosogyrous than *I. ginterensis*. *I. beringensis* also has a much more prominent projecting beak and umbo, a smaller posterior auricle, and a well-defined auricular sulcus.

The recognition of *I. ginterensis* in the Western Interior of North America provides yet another close biostratigraphic tie between the Pacific and Atlantic faunas.

Stratigraphic and Geographic Position. *Inoceramus ginterensis* is abundant in units 2 through 24 of the lower and lower middle parts of the Hartland Member, section 3; it is much less abundant and less well preserved in units 52 through 63 of the upper middle and upper parts of the Hartland Member, regional zone of *Dunveganoceras albertense*, in thin-bedded to blocky chalk, shaly chalk, and chalky limestone, primarily as whole single valves or open, semiarticulated bivalved specimens. Questionably related fragments occur in units 74 and 76, section 2, in the uppermost part of the Hartland Member and lowest part of the Bridge Creek Member.

Types. Illustrated hypotypes: right valve with hinge structures preserved—USNM 169347 (unit 4, sec. 3; Pl. 5, fig. 3; Pl. 4, fig. 1); nearly complete internal molds of adult right valves—USNM 169348 (unit 4, sec. 3; Pl. 4, fig. 4), USNM 169349 (unit 4, sec. 3; Pl. 4, fig. 3), USNM 169350 (unit 24, sec. 3; Pl. 3, fig. 3); small right valve, internal mold, showing juvenile ornament well—USNM 169351 (unit 4, sec. 3; Pl. 3, fig. 2); immature right valve with juvenile ornament—USNM 169352 (unit 4, sec. 3; Pl. 1, fig. 2); nearly complete internal molds of adult left valves—USNM 169353 (unit 4, sec. 3; Pl. 3, fig. 4), USNM 169354 (unit 4, sec. 3; Pl. 4, fig. 6), USNM 169355 (unit 4, sec. 3; Pl. 4, fig. 5), USNM 169356 (unit 4, sec. 3; Pl. 5, fig. 2), USNM 169357 (unit 4, sec. 3; Pl. 5, fig. 5), USNM 169358 (unit 3, sec. 3; Pl. 3, fig. 6). A pair of associated immature valves, USNM 169359 (unit 4, sec. 3; Pl. 3, fig. 1). Unillustrated, measured, adult right valves, USNM 169350-169363 (unit 4, sec. 3); measured adult left valves, USNM 169364-169366 (unit 4, sec. 3). Unmeasured and unillustrated specimens from section 3: USNM 169367 (unit 7), USNM 169368 (unit 9), USNM 169369 (unit 11), USNM 169370 (unit 13), USNM 169371 (unit 15), USNM 193497 (unit 19), USNM 169372 (unit 21), USNM 169373 (unit 22), USNM 169374 (unit 24). Unmeasured and unillustrated specimens from section 2: USNM 169375 (unit 50), USNM 169376 (unit 52), USNM 169377 (unit 54), USNM 169378 (unit 56), USNM 169379 (unit 58), USNM 169382 (unit 59), USNM 193498 (unit 61), USNM 169380 (unit 62), USNM 169381 (unit 63).

Inoceramus tenuistriatus Lineage

Inoceramus tenuistriatus, as broadly defined by Nagao and Matsumoto (1939), represents a Cenomanian-Turonian lineage that is characterized by the following:

subquadrate to nearly square shell outline; moderate biconvexity with the left valve being somewhat more convex than the right and having a more projecting, inflated, moderately to strongly prosogyrous beak and umbo; slightly concave anterior and dorsoposterior margins and a moderately rounded, subevenly convex ventral margin; a broad but prominent umbonal fold separated from a well-defined, triangular, moderate-sized posterior auricle by a shallow auricular sulcus; and ornamentation consisting typically of weak to moderately developed, closely to distantly and subevenly spaced asymmetrical rugae, steepest ventrally, between and over which are moderate-sized to very fine, raised, equally to subequally developed growth lines or slightly raised lamellae.

The *I. tenuistriatus* lineage seems to arise during the Middle Cenomanian from the *Inoceramus pictus* lineage, to which the lineage was previously assigned. The subquadrate shell outline, well-defined posterior auricle, and the importance of regularly spaced rugae in the ornament seem to distinguish consistently the component species of the *I. tenuistriatus* lineage from those of the *I. pictus* lineage. In the Western Interior the lineage is poorly known and secondary in abundance and diversity to the *I. pictus* lineage. *Inoceramus tenuistriatus*? Nagao and Matsumoto and *I.* sp. aff. *I. tenuis* Mantell (s. l.) are the only Oklahoma representatives and are found elsewhere in the Western Interior along with two to three additional but as yet unidentified Cenomanian species. In Europe and Japan, however, the lineage is extremely important, and a full succession of Cenomanian-Turonian species is preserved. In the Cenomanian strata of Europe these are primarily forms like the quadrate variants of *I. tenuis* Mantell (s. l.), *I. etheridgei* Woods, *I. conicus* Gueranger, and others. The widespread distribution of the *I. tenuistriatus* lineage, which is basically cosmopolitan, and the local abundance of taxa belonging to it suggest that when studied in the Western Interior, it will have considerable importance in regional biostratigraphic zonation and correlation.

Inoceramus tenuistriatus? Nagao and Matsumoto (Pl. 2, fig. 5; Pl. 3, figs. 7, 9, 10)

Inoceramus tenuistriatus Nagao and Matsumoto, 1939, Pl. 24, figs. 3, 8a–8c; Pl. 26, figs. 1a–1c, 2a, 2b; non Pl. 26, figs. 3, 4. Pergament, 1966, p. 47, 48, Pl. 14, figs. 2–4.

Inoceramus pictus neocaledonicus Jeannet. Pergament, 1966, p. 55–57 (part), Pl. 30, figs. 2, 3(?); non Pl. 30, fig. 4; Pl. 29, figs. 3(?), 4(?); non Pl. 33, fig. 1.

?*Inoceramus pictus* Sowerby. Sornay, 1965, p. 9 (part), Figs. 4, 5. Probably within many authors' concept of this species.

Material. Two moderately well preserved internal molds of right valves, small- and moderate-sized adults, with broken posterior flanks. An internal mold of an adult left valve; beak crushed and posterior flank broken. Internal mold of a complete but crushed juvenile left valve.

Description. Small to moderate size for genus attaining maximum height of 50 mm; Table 4 gives summary of measurements. Moderately prosocline, average angle of inclination 56.5° (right valve), 64° (left valve); inequivalve and moderately biconvex; left valve significantly more convex and dorsoanteriorly projecting (beak-umbo) than right (compare Pl. 2, fig. 5, with Pl. 3, fig. 9). Maximum convexity dorsocentral and anterior to midline, near base of umbo. Anterior flank steeply inclined to vertical and slightly concave; weakly defined lunule developed, bounded by low, rounded rim. Ventral, ventroposterior flanks moderately to gently sloping;

Table 4. Summary of measurements for *Inoceramus tenuistriatus*? Nagao and Matsumoto

Character*	USNM 169402	USNM 169403	USNM 169399	USNM 169401
Valve	Right	Right	Left	Left
H	42 mm	28 mm	50 mm	17.5 mm
L	39 mm†	25 mm†	44 mm†	18 mm
W	10.5 mm	5 mm	15 mm	6 mm
LHL	23 mm†	11 mm†	20 mm†	11 mm†
a	95°	90°	83°	98°
i	56°	57°	67°	61°
R/AL_1	4.5	3	5	2
R/AL_2	5.5	. .	4	. .
GL/AL_1	. .	42	38	37†
GL/AL_2	15	. .	16	. .

*Character measurement explained in Figure 1. R = number of rugae. GL = number of growth lines. AL_1 = first 25 mm of axial length. AL_2 = second 25 mm of axial length.
†Reconstructed.

dorsoposterior flank gently sloping to flattened on posterior auricle. Auricle subtriangular, poorly defined, separated from disc by moderate change in postumbonal slope. Strongly prosogyrous (Pl. 3, figs. 7, 9); beak blunt, umbo strongly (left valve) to moderately (right valve) inflated. Beak-umbo slightly (right valve) to moderately (left valve) projecting above hinge line, markedly incurved. Shell outline subquadrate (Pl. 3, fig. 9); height only slightly greater than length. Dorsal margin long, straight; anterior margin straight to slightly concave, curvature of posterior and ventral margins slightly convex.

Juvenile ornament consisting of very fine, subequally developed, evenly and closely spaced, raised rounded growth lines over much of umbo (Pl. 3, figs. 7, 10). A prominent ruga separates juvenile from adult shell portions. Adult ornament characterized by equally fine, but more irregularly developed growth lines, becoming fainter and somewhat larger with age, and numerous, strong, moderately and subevenly spaced rugae (Pl. 3, figs. 9, 10).

Only prismatic shell layer poorly preserved, moderately thin over most of valve, only slightly thickened near ligamental plate (not fully preserved). Musculature poorly preserved (Fig. 3). Pallial line entire, very thin, faint anteriorly where preserved. Posterior adductor muscle track faint, moderately wide, situated on interior of umbonal fold. Entire inner shell surface covered by moderately strong, closely spaced, narrow grooves, the trace of mantle retractor muscles.

Remarks. The poor state of preservation in most Oklahoma specimens, their small number, and the slight morphologic differences between them and *Inoceramus tenuistriatus* from Japan cause us to question the species identification at present. These morphologic distinctions are within the expected range of variation for inoceramid populations, yet seem to be consistent between the two areas. Oklahoma specimens are larger, more regularly and coarsely rugate, and apparently slightly more inequivalve than Japanese specimens illustrated by Nagao and Matsumoto (1939, Pl. 24, figs. 3, 8a-8c, Pl. 26, figs. 1-4) and Pergament (1966, Pl. 13, fig. 2, Pl. 14, figs. 2-4). Some Japanese representatives (Nagao and Matsumoto, 1939, Pl. 26, figs. 3, 4) have somewhat coarser juvenile growth lines. Nevertheless, the Western Interior and Indo-Pacific forms are very closely related if not conspecific and represent another significant biostratigraphic tie between these distant areas. The Indo-Pacific forms are reported from Middle Cenomanian through Turonian rocks, and thus overlap the range of *I. tenuistriatus*? in the Western Interior United States. Pergament's specimens of *I. pictus neocaledonicus*, here referred to this species (1966, Pl. 27, figs. 3, 4(?), Pl. 30, fig. 2), are also of Late Cenomanian age and seemingly occur in this substage with *I. flavus pictoides* Sornay (Pergament, 1966, Pl. 30, fig. 4). The latter also occurs in the Cenomanian strata of Oklahoma, but has a slightly lower stratigraphic range than *I. tenuistriatus*? (see Fig. 5 in Pt. 1 of this paper).

Stratigraphic and Geographic Occurrence. Units 58 and 62 (middle part of Hartland Member), unit 74 (upper part of Hartland Member), and units 78 and 81 (lower part of Bridge Creek Limestone Member), section 2, Greenhorn Formation. A poorly preserved specimen possibly related to *I. tenuistriatus* was found in unit 86, section 2. All specimens are from platy to slabby, moderately resistant, tan chalky limestone beds.

Figured Specimens. Largest adult right valve, USNM 169402 (unit 81, sec. 2); largest adult left valve, USNM 169399 (unit 81, sec. 2); small adult right valve, USNM 169403 (unit 58, sec. 2); immature left valve, USNM 169401 (unit 62, sec. 2). All specimens internal molds.

Inoceramus sp. aff. *I. tenuis* Mantell, s. l.
(Pl. 2, fig. 2; Pl. 3, figs. 5, 8)

Refer to *I. tenuis* Mantell, 1822, p. 132. Woods, 1911, p. 271-273, Pl. 48, fig. 1 especially.

Material. An adult right valve, internal mold, beak and posterior auricle poorly preserved. Internal mold, nearly complete dorsal half of adult left valve. One well-preserved juvenile right valve with crushed beak.

Description. Moderate size, attaining height of 51 mm, length 42 mm, width (right valve) 13 mm; moderately prosocline, average angle of inclination 55°. Moderately biconvex, slightly inequivalve, left valve somewhat more inflated, especially umbonally, than right. Maximum inflation dorsocentral and anterior to midline, at base of umbo. Anterior flank nearly vertical, without distinct lunule; other flanks moderately to slightly inclined. Posterior auricle small, flattened, subtriangular, separated from disc by strong break in postumbonal slope. Prosogyrous; beak blunt, rounded; umbones inflated, incurved, poorly differentiated from main part of disc (Pl. 3, fig. 8). Outline subovate (Pl. 3, figs. 5, 8); anterior, posterior margins slightly and convexly curved; ventral margin moderately curved; posterodorsal margin (hinge line) short, straight. Apical angle slightly obtuse.

Ornament similar on late juvenile and adult shell, consisting of subequally developed, moderately spaced, flat to slightly raised growth lines and lamellae (Pl. 3, fig. 8), closer and more prominent dorsally. Early juvenile shell (Pl. 3, fig. 5) with very fine, close, equal lamellae. Lamellae of adult shell superimposed on moderately and subevenly spaced, shallow asymmetrical rugae (Pl. 3, fig. 8); 2 rugae, 18 growth lines in first 25 mm of axial length; 2.5 rugae, 13 growth lines in second 25 mm of axial length.

Ligamental plate moderately thickened, but resilifers unknown. Pallial line very thin, faint, entire, incompletely known (Fig. 3). Posterior adductor insertion area shallow, subcrescentric. Faint radial grooves on shell interior (Fig. 3) from tracks of mantle retractors. No pedalbyssal musculature observed.

Remarks. The specimens available for study are too poorly preserved for definite identification or extensive description. In shape, beak-umbo characteristics, relative valve convexity, and nature of the growth lamellae, these forms closely resemble earlier variants of *I. tenuis* Mantell as broadly defined by Woods (1911, p. 271-273) and especially as illustrated from the Red Limestone at Louth, England (Woods, 1911, Pl. 48, fig. 1). These variants differ from typical *I. tenuis* (Woods, 1911,

Figs. 31, 32) in having a less projecting, broader, less convex beak-umbo on the left valve, closer growth lamellae, and in being more erect, with less relative height. Ultimately these two forms should be split into distinct taxa. Oklahoma specimens related to *I. tenuis* are distinct in having coarser juvenile ornament and distinct, subregular rugae over most of the shell. *Inoceramus etheridgei* Woods (Cenomanian) is very similar in general shape and the nature of the constrictions but has much finer lamellae and a concave anterior margin. *Inoceramus scalprum* Boehm is probably related but is distinct in having finer lamellae and lacking prominent rugae. *I. tenuistriatus* Nagao and Matsumoto (Upper Cenomanian–Turonian) is also similar in form and development of rugae but differs in having coarse, raised growth lines over the entire valve.

Stratigraphic and Geographic Position. Specimens of *Inoceramus* sp. aff. *I. tenuis* Mantell s. l. are known from units 20, 22, 24, and 26 (questionably this form), section 2, in the lower part of the Lincoln Member, Greenhorn Formation, regional zone of *Dunveganoceras pondi*.

Figured Specimens. Adult right valve, internal mold, USNM 169405 (unit 22, sec. 2). Dorsal part of adult left valve, interior of shell and latex cast, USNM 169404a, 169404b (unit 24, sec. 2). Juvenile right valve, interior of shell and latex cast, USNM 169398a, 169398b (unit 24, sec. 2).

Inoceramus rutherfordi Lineage

Though poorly represented in the Oklahoma sequence, the *Inoceramus rutherfordi* lineage is an important Middle Cenomanian stock in the Western Interior and Gulf Coast, with considerable biostratigraphic potential. Two species are described and are moderately common in North America: *Inoceramus rutherfordi* Warren and *I. arvanus* Stephenson.

The lineage is characterized by having medium-sized, subquadrate, moderately inequivalve, prosocline shells with strong posteroventrally directed folds and sulci. The left valve is more inflated than the right, especially the left beak-umbonal area which projects above the hinge line and overhangs the beak of the right valve. Ornament consists of moderately low, rounded, subregularly spaced rugae between and over which are faint, flat to slightly raised, subregularly to irregularly developed growth lines and (or) lamellae. The posterior adductor muscle (noted on *I. arvanus*) is comma-shaped and bears a dorsal extension that is probably a large pedalbyssal muscle scar, as in *Isognomon*. This is unique among the Inoceramidae as far as is known, and in combination with the strong posterior folds and sulcus, these characteristics possibly distinguish the group generically or subgenerically from *Inoceramus* s. s.

Systematic study of the lineage by utilizing stratigraphically closely spaced populations with known temporal relationships demonstrates that both named "species" are plexus of two or more taxa, mainly distinguished by evolutionary changes in the folds, sulci, and ornamentation. These are not yet named, but Kauffman (1966a) has already used them biostratigraphically (as early form, late form, etc.) within the *I. rutherfordi* plexus to zone the Middle Cenomanian. At least two species seem to exist in the present concept of *I. arvanus* as well.

Unfortunately, the Oklahoma specimens are too poorly preserved to determine which of these stratigraphically distinct forms are present.

Inoceramus? *rutherfordi*? Warren, s. l.

Inoceramus rutherfordi Warren, 1930, p. 59, Pl. 7, figs. 1-3. McLearn, 1945, Pl. 3, fig. 1. Hattin, 1965a, p. 50, Pl. 4, figs. C, E, F, H.

Material. A single incomplete small adult right valve with portions of the posterior flank, beak, and posterior auricle missing. Internal mold with some shell preserved. Additional fragments. Description based on this material and stratigraphically equivalent specimens in the Western Interior.

Description. Moderately biconvex, left valve slightly more inflated, maximum inflation on the ventral umbo; anterior flank very steep, with prominent, slightly concave lunule having a raised rounded marginal rim; ventral and posterior flanks moderately to gently sloping; posterior auricle flattened, triangular, small to medium sized, separated from disc by a prominent auricular sulcus. Umbonal fold prominent, moderately arched, asymmetrical, with shallow sulcus posterior to it; posterior sulcus separated from auricular sulcus by a strong postumbonal fold (see Warren, 1930, Pl. 7, figs. 1, 3). Sulci and folds very weakly developed or missing on young shells (the Oklahoma specimen). Anterior "auricle" or marginal salient developed on adult valves below beak, as extension of inner edge of lunule. Moderately prosocline and prosogyrous; beaks rounded, blunt; beak-umbo moderately projecting above straight dorsal margin, more so on left valve than right. Outline elongate-ovate in young, subquadrate in adult with strong development of auricles; anterior, posterior commissure straight to very slightly curved, ventral margin slightly curved, with shallow re-entrants in outline at intersection with sulci. Ornament of juveniles and young adults (Oklahoma specimen) consisting of numerous very fine, closely and unevenly spaced, weakly developed growth lines and in some cases very faint small rugae. Adults have similar growth lines, a few stronger ones, and somewhat more prominent, moderately and unevenly spaced rugae. Shell moderately thin. Hinge characters as on *I. arvanus* Stephenson (1952, Pl. 12, figs. 8, 9), with thickened ligamental plate and numerous, moderately concave resilifers separated by thinner flat-topped platforms. Musculature unknown.

Remarks. The young adult specimen from Oklahoma had not yet developed the adult characters that allow species differentiation within the *I. rutherfordi* plexus, and thus the species assignment must be questioned even though *I. rutherfordi* s. s. has previously been reported and is common from this level in the Western Interior (see, for example, Hattin, 1965a). The early growth characters clearly place it within the plexus, however. The assignment to *Inoceramus* is questioned because musculature characters of *Inoceramus arvanus* Stephenson, the possible ancestor of *I. rutherfordi*, are clearly distinct from those of true *Inoceramus* (*I. cuvieri*, *I. lamarcki*) in having a strong pedalbyssal muscle adjacent to the dorsal part of the posterior adductor as in *Isognomon*. This may reflect generic or even higher taxonomic differentiation.

Geographic and Stratigraphic Occurrence. Lower part of upper shale member of the Graneros Shale (units 6 and 8, sec. 2); precise biostratigraphic zone unknown.

Described Specimen. USNM 172867.

Genus *Mytiloides* Brongniart, 1822

Mytiloides Brongniart, 1822, p. 320. Conrad, 1874 (1875), nom. null. (fide Cox, 1969). Cox, 1969, p. N317.
Mytiloida, Brongniart, 1832 (nom. van., fide Cox, 1969, p. N317).
Orpheoceramus Heinz, 1932, p. 304 (nom. nud., fide Cox, 1969, p. N317).
?*Citharoceramus* Heinz, 1932, p. 13 (nom. nud., fide Cox, 1969, p. N317).
?*Aristoceramus* Heinz, 1932, p. 258 (nom. nud., fide Cox, 1969, p. N317).

Type Species. *Ostracites labiatus* Schlotheim, 1813, monotypy.

Description. Subequivalve, strongly to moderately prosocline; outline subovate to elongate-ovate (labiatoid); slightly biconvex; very thin shells. Beak-umbo anterior and slightly to moderately projecting, suberect to prosogyrous, commonly with distinct juvenile ornamentation. Small anterior auricle or marginal projection rarely present; posterior auricle small to moderate size, subtriangular, flattened, normally without prominent auricular sulcus; hinge line of moderate length. Ornament normally of small to moderate size, asymmetrical concentric rugae, alternating with numerous, crowded, regularly to irregularly developed, small, raised growth lines and (or) flat lamellae; concentric constrictions and shallow posteroventrally directed sulci rarely developed; distinct growth lines absent on some species. Hinge area only slightly thickened, with small, closely spaced, multivincular resilifers; edentulous. Musculature simple, insertion areas weakly impressed, consisting of entire or nearly entire pallial line; and a narrow, laterally elongated posterior adductor area; no pedalbyssal musculature or apparent byssal gape in shell.

Discussion. Cox (1969) and all previous authors undertaking generic diagnoses of Inoceramidae have almost totally ignored interior shell morphology in distinguishing genera and higher taxa, in particular musculature and to a lesser extent dorsal structures of the hinge area (for example, nature of the ligamental area, development of pseudodentition, nature of the byssal slit where present, and internal ribs of the dorsal shell area). These features are not commonly well preserved but are biologically of the utmost importance and occur on sufficient specimens to warrant their careful consideration in classification. Kauffman (1966b) summarized their importance in this respect and outlined the main groupings of inoceramids based on interior shell morphology. These groupings crosscut all existing classifications based on exterior shell characters, and this is attributed to widespread external homeomorphy among distinct groups of inoceramids adapting to similar environmental situations.

In this generic study it has become obvious that *Mytiloides* is a distinct genus of inoceramids not closely related to *Inoceramus* s. s. The differences between them are so great that placing *Mytiloides* as a subgenus of *Inoceramus* s. s. (Cox, 1969, p. N317) is not warranted. Externally they are distinguished by the erect form, moderately to grossly inflated, inequivalve shells, commonly strong sulci, folds, and auricles, and distinct ornament pattern of *Inoceramus* s. s. (alternating rugae and even, flat lamellae). Internally the distinctions are also considerable. *Inoceramus* has a massive hinge with medium- to large-sized resilifers, a prominent sinuous dorsoanterior byssal slit, generally an entire to partially pitted pallial line, a larger insertion area for the comma-shaped posterior adductor muscle, and in some cases, prominent to weakly defined, small pedalbyssal muscle insertion area(s) and strong pallial (mantle) retractor tracks. *Mytiloides* has no pedalbyssal muscles, no obvious byssal slit, a continuous or nearly continuous pallial line without strong mantle retractor tracks, and an elongate, subcrescentric posterior adductor muscle insertion area. The hinge plate is thin, with small multivincular resilifers, and is not extended anterior of the beaks.

Mytiloides labiatus (Schlotheim) Lineage

Mytiloides "*labiatus*" (Schlotheim) ("*Inoceramus labiatus*" of most previous authors) is one of the best known and most cosmopolitan of the Inoceramidae, being widely reported in rocks of Late Cenomanian and Turonian age throughout the Americas and Europe (including Tethyan occurrences) and less commonly in Africa and the Indo-Pacific. Miscellaneous faunal lists and various museum collections that we have examined also ascribe the "species" less commonly to Coniacian–Lower Santonian rocks and rarely (and probably incorrectly) to rocks of Albian, Early Cenomanian, and Middle Santonian through Campanian age.

Several authors (for example, Simionescu, 1899; Böse, 1923; Seitz, 1934 [1935]) have recognized that the "species" *I. labiatus*, as commonly used in the literature and identified in collections over much of the world, actually represents a rapidly evolving lineage composed of numerous species and subspecies ranging in age from Latest Cenomanian through the Coniacian stages. But none of these authors has suggested the evolutionary history of the lineage and the biostratigraphic distribution of the component species. The stratigraphic distribution of Oklahoma taxa within the lineage provides valuable information relative to its evolutionary history, although the record is far from complete.

Schlotheim's original description of the species (1813, p. 93, as *Ostracites labiatus*) is vague, and hard to obtain, and the types are now apparently lost (Seitz, 1934 [1935], p. 450). Subsequently, the generally accepted concept of the species has become that of *M. mytiloides* (Mantell), characterized by curved, elongated, highly inclined, thin, subequal valves with slight to moderate inflation, a triangular posterior auricle, terminal anterior beaks, and a characteristic ornament consisting of alternating, regularly spaced concentric rugae and moderate to fine, slightly raised, regularly spaced growth lines (Pl. 6, figs. 13, 14, 16) (for example, the concept of Stanton, 1893 [1894], p. 77, the last, highly definitive description of the species in America). *Mytiloides mytiloides* is of Early Turonian age. But, as variously used in the world literature, the "species" *M. labiatus* has no consistency in concept or precise stratigraphic range, and considerable morphologic variation in form and arrangement of ornament is attributed to it.

Seitz (1934 [1935]), whose long experience with the German Inoceramidae gives him authority in the matter, pointed out that *M. labiatus* as originally defined by Schlotheim is *not* synonymous with *M. mytiloides* (Mantell), but rather is characterized by having a much more inflated, umbonally curved shell with irregular, somewhat crowded rugae and numerous, fine, flat-lying lamellae. The beak and umbo are very swollen, suberect, and moderately projecting above the hinge line (Seitz, 1934 [1935], Figs. 5, 6; Pl. 38, fig. 1, which Seitz [July, 1971, oral commun.] said is closest to the type specimen). This is in marked contrast to the regular ornament, low convexity, small beak, and more inclined but less curved shell of *M. mytiloides* (Seitz, 1934 [1935], Pl. 36, figs. 2, 4). Seitz considered *M. labiatus* and *M. mytiloides* as distinct "varieties" (subspecies) of *M. labiatus*. Thus, the generally accepted American concept of *M.* "*labiatus*" is actually that of *M. labiatus mytiloides* (Mantell) of Seitz; we choose to accept *M. mytiloides* as a distinct species. *Mytiloides labiatus labiatus* s. s. has been recorded from American lower Middle Turonian rocks. Seitz (1934 [1935]) further recognized the "varieties" (used as subspecies) *M. subhercynica* (Seitz), *M. opalensis* (Böse), and *M. submytiloides* (Seitz) within the lineage, but did not clarify their stratigraphic distribution or state whether they represent variation within a single population from one isochronous surface or variation through time or both.

An extensive systematic revision of the *M. labiatus* lineage from the Americas and Western Europe is under way by Kauffman. The first phase of this has been a biometric analysis of large, stratigraphically closely spaced populations ascribed to *M.* "*labiatus*" from Upper Cenomanian through Lower Coniacian rocks of the Western Interior United States. Whereas this has not yet been completed, sufficient evidence is available (Kauffman, 1970, Fig. 4) to suggest that several biostratigraphically useful species and subspecies originally included within the concept of *M.* "*labiatus*" (s. l.) make up the lineage in this area. Principal characters useful in systematically splitting the lineage are valve form, convexity, inclination, nature of the posterior auricle, presence or absence of an anterior auricle, development and projection of the beak-umbo, and density, strength, and regularity of the concentric ornamentation. Interior shell characters and shell structure remain fairly uniform.

In Cimarron County, Oklahoma, exposures of the Bridge Creek Limestone Member, Greenhorn Formation, are commonly weathered, highly fractured, and incomplete. Well-preserved specimens belonging to the *M. labiatus* lineage are not common, and statistically significant populations are not available. This material therefore is not being incorporated directly into the main population systematic study of the *M. labiatus* lineage in the Western Interior and can only be treated typologically—that is, referred to species or subspecies whose typical forms our few specimens from Oklahoma resemble most closely.

Open nomenclature is applied in cases where new taxa are recognized but are to be described from population analysis of much larger collections at a later date. The Oklahoma forms do seem to reflect, however, the early evolutionary progression of characters, species, and subspecies within the lineage that can be recognized in more extensive population analyses of other Western Interior collections.

Five distinct forms characterize the *M. labiatus* lineage in the Bridge Creek Member and its lower beds that are transitional into the Hartland Member, Greenhorn Formation, of Cimarron County (Fig. 4). These are, from lowest to highest:

(1) *Mytiloides* sp. aff. *M. submytiloides* (Seitz) (Pl. 6, figs. 7, 10), from the lowest bed of the Bridge Creek Limestone Member and upper part of the Hartland Member (Upper Cenomanian); (2-4) *Mytiloides opalensis* (Böse) (Pl. 6, figs. 3, 6), *M.* sp. transitional between *M. opalensis* and *M. mytiloides* (Pl. 6, fig. 4), and *M.* sp. aff. *M. duplicostatus* (Pl. 6, fig. 5), all from the middle part of the exposed Bridge Creek Limestone Member; and (5) *Mytiloides mytiloides* (Mantell) from the upper part of the exposed portion of the Bridge Creek Limestone Member (Pl. 6, figs. 13, 14, 16).

General evolutionary trends within this part of the lineage are toward increasing size, valve inclination, inflation, increased projection and obliquity of the beak-umbo, and greater regularity of concentric ornament. These early evolutionary trends within the lineage, shown by population analysis of Western Interior members of the lineage from the Bridge Creek of Colorado and Kansas, are reflected in the small collections from Cimarron County. In younger rocks the lineage diversifies into two lines of descent, (1) obliquely inclined, finely and (or) irregularly ornamented forms (for example, *M. labiatus hercynicus*, *M.* "*problematicus*" of American usage) and (2) more rounded, less inclined, regularly ornate forms (for example, *M.* "*latus*" in the sense of Hattin, 1962). Both lines survived to Late Turonian time in the Western Interior, but only the more oblique-shelled group is known to occur in younger strata.

***Mytiloides mytiloides* (Mantell)**
(Pl. 6, figs. 11-16)

Inoceramus mytiloides Mantell, 1822, p. 215, Pl. 27, fig. 3; Pl. 28, fig. 2.

Note. The extensive synonymy below attempts to incorporate most of the works that we have been able to examine that wholly or partially include the concept of *M. mytiloides* presented here. A (?) indicates that the description and (or) illustration is not definitive enough to be certain; (part) means that other taxa are also included in a broad species concept along with *M. mytiloides* (s. s.).

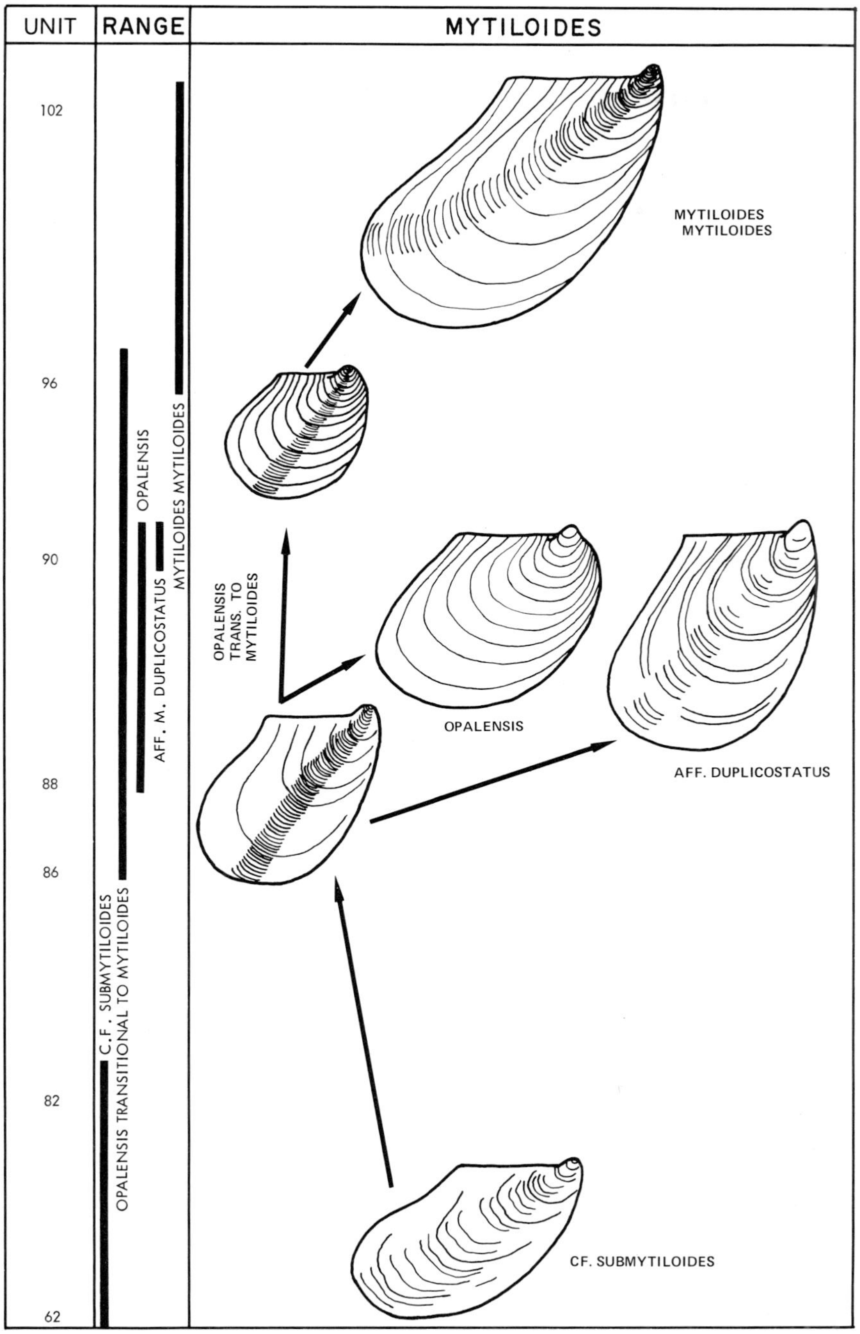

Figure 4. Stratigraphic distribution and principal morphologic features of taxa within the *Mytiloides labiatus* lineage (diagrammatic; composite sketches from several specimens). All from section 2 (see Fig. 3 in Part 1). Arrows suggest evolutionary relationships.

Inoceramus mytiloides Sowerby, 1825, p. 62, Pl. 442. Goldfuss, 1836, p. 118 (part), Pl. 113, fig. 4. Bronn, 1853-1856, p. 289, 290 (part), Pl. 32, fig. 10. Morris, 1854, p. 17 (part). Strombeck, 1863, p. 119-121 (part).

Catillus mytiloides Pusch, 1837, p. 45 (part).

Inoceramus problematicus d'Orbigny, 1843, p. 510 (part), Pl. 406, Figs. 1(?), 5-7, non figs. 2-4. Meek and Hayden, 1857, p. 14 (part). Pictet, 1868-1871, p. 113, 114 (part). Meek, 1872, p. 435-461, (part; bed 3 at Coalville, non(?) bed 7 at Coalville; non bed 12 at Bear River City, Wyoming). Meek, 1876, p. 46 (62) (part), Pl. 9, fig. 3b. Hill, 1889, p. 9 (part).

Inoceramus pseudo-mytiloides Schiel, 1855, p. 108 (part), Pl. 3, fig. 8.

Inoceramus labiatus Stoliczka, 1869 (1871), p. 408, Pl. 29, figs. 1, 1a. Geinitz, 1872-1875, p. 46-49 (part), Pl. 12, figs. 2, 3(?) (non fig. 1). Geinitz, 1873, p. 13, 14 (part). Schluter, 1877, p. 262, 263. Dixon and Jones, 1878, p. 215, Pl. 63, fig. 3; Pl. 64, fig. 2. Stanton, 1893 (1894), p. 77, 78 (part), Pl. 10, fig. 4; Pl. 14, fig. 2(?). Kayser, 1893, Pl. 55, fig. 4 (= *mytiloides*). Gilbert, 1896, Pl. 57. Credner, 1897, p. 633, Fig. 493. Logan, 1898, p. 450 (part), Pl. 92, fig. 4. Simionescu, 1899, p. 23 (part), Pl. 1, fig. 9(?); Pl. 2, fig. 2(?). Hill, 1901 (part), Pl. 40, fig. 1(?) (non fig. 4). Darton, 1905, Pl. 24, fig. B. Darton, 1906, Pl. 1, fig. B. Veatch, 1907, Pl. 9, fig. 3a(?). Darton, 1909, Pl. 12, fig. B. Grabau and Shimer, 1909, p. 443, Fig. 583(?). Böhm, 1909, p. 117 (part). Böse, 1910, p. 657(?), 658(?). Fraas, 1910, Pl. 38, fig. 9(?). Woods, 1911, p. 281-284 (part), Fig. 37(?), Pl. 50, figs. 1-4. Böse, 1913, p. 25-28 (part), Pl. 2, figs. 3, 6(?), non figs. 1, 2, 4, 5. Reeside, 1923, p. 31 (part), Pl. 14, figs, 3, 5, non fig. 4. Dietrich, 1924, p. 87(?). Kayser, 1924, Pl. 23, fig. 5. Heinz, 1928, p. 61. Adkins, 1928, p. 94 (part; middle Eagle Ford Formation). Heine, 1929, p. 56 (part). Seitz, 1934 (1935), p. 430-473 (part), Figs. 2a-2f (as *I. labiatus mytiloides*), 9c (non 9a, 9b), 11c(?) (as *I. labiatus labiata*). Bertin, 1939, Fig. 380b. Moreman, 1942, p. 199(?), 200(?) (part), non Pl. 31, fig. 4. Gryc, 1951 (1952), Fig. 20. Rutsch and Salvador, 1954, Pl. 40, fig. 1. Shimer and Shrock, 1955, p. 389 (part), Pl. 151, fig. 15(?). Cobban, 1958b, p. 118(?). Ronchetti, 1961, Pl. 26, fig. 2(?), non figs, 1, 3, 4. Haller, 1963, p. 158(?). Hattin, 1965b, p. 15, 16, Fig. 3, no. 10. Sornay, 1966, p. 5-36 (part), Figs. 2, 5.

Inoceramus mytilopsis Conrad, 1857, p. 152 (part), Pl. 5, fig. 6b (non figs. 5, 6a).

Mytiloides labiatus Heinz, 1933 (1934), p. 248, 249 (part), Pl. 17, fig. 2, 3(?), non fig. 12.

Inoceramus labiatus cedroensis Maury, 1936, p. 113 (part), Pl. 8, fig. 17, non fig. 12.

Inoceramus (*Mytiloides*) *labiatus*, Dacqué, 1939, p. 103, Pl. 5, figs. 4(?) 5(?); Pl. 6, figs. 11(?)-13(?) (part). Rutsch and Salvador, 1954, p. 419, Pl. 40, fig. 1. Jones and Gryc, 1960, p. 160, 191 (part), Pl. 19, figs. 1, 3(?), 5(?).

Inoceramus glennensis Anderson, 1958, p. 99, 100, Pl. 17, figs. 1, 2.

?*Inoceramus paramytiloides* Sornay, 1965, p. 13, 14, Figs. 6, 7, Pl. C, figs. 1-4.

Inoceramus labiatus var. *mytiloides*, Seitz, 1934 (1935), p. 435-444 (part), Figs. 2a-2f.

Inoceramus labiatus var. labiata Seitz, 1934 (1935), p. 448-454 (part), Figs. 9c (non 9a, 9b), 11c(?).

Material. Ten nearly complete internal molds of juvenile (3) and adult (7) left (5) and right (5) valves preserving the ornament and major features of the shell. Numerous additional large fragments.

Table 5. Summary of Measurements for *Mytiloides mytiloides* (Mantell)

Character*	No. of specimens	Range	Average
H	9	15–58 mm	31 mm
L	9	17–80 mm	34.2 mm
W	9	4–17 mm	8.25 mm
LHL	8	6–30 mm	15.4 mm
a	5	79–113°	96.2°
i	6	40–59°	42°
R/AL_j	6	11–14	12.5
R/AL_1	5	13–20	16.4
R/AL_2	4	7–9	7.8
R/AL_3	2	4–5	4.5
R/AL_4	1	3	3

*Character measurements explained in Figure 1. R = number of rugae. AL_j = axial length of juvenile part of shell (that is, the first 15 mm). AL_1 = first 25 mm of axial length. AL_2 = second 25 mm of axial length. AL_3 = third 25 mm of axial length. AL_4 = fourth 25 mm of axial length.

Description. Shell attaining large size for genus, maximum height 65 mm, length 73 mm in Oklahoma specimens; shells twice this size known elsewhere in Western Interior. Measurements summarized in Table 5. Subequivalve, moderately biconvex, maximum convexity dorsocentral, at base of umbo. Anterior flank and lateral umbonal slopes moderately to very steep, other flanks flattened or slightly inclined to umbo. Posterior auricle moderate sized, flattened, subtriangular, separated from disc by marked break in postumbonal slope, without well-defined sulcus. Shell strongly prosocline, average angle of inclination 42° (Pl. 6, figs, 13, 14). Beaks bluntly pointed, anterior, terminal, projecting moderately above hinge line, slightly incurved; beak-umbo strongly prosogyrous in adult shells. Umbonal fold asymmetrically rounded, broad, poorly defined ventrally. Valve outline inclined, tongue-shaped (labiatoid) to obliquely elongate-ovate (Pl. 6, figs. 11–16). Hinge line straight, of moderate length for genus, equaling one-third to one-half total shell length; anterior, dorsoposterior margins straight to slightly curved (convex outward), other margins moderately, asymmetrically rounded.

Juvenile and adult ornamentation distinct. Umbonal (juvenile) ornamentation to 15 or 20 mm of axial length consisting of small, closely and subevenly spaced, subequally developed, low to moderately strong rugae with or (more rarely) without numerous fine, crowded, flattened growth lines irregularly developed on or between rugae (Pl. 6, figs. 11, 12, 15). Junction between adult and juvenile ornament moderately distinct. Adult ornament characteristically consisting of subevenly developed, prominent, moderately and evenly spaced, asymmetrical rugae with steep distal flanks, between and on which lie evenly spaced, subequally developed, strongly to moderately raised growth lines, characteristically 3 to 5 between adjacent rugae (Pl. 6, fig. 16). In some specimens, adult ornament more irregular on umbo near junction with juvenile ornament (Pl. 6, fig. 13) or on distal portions of old shells (a girontic feature).

Ontogenetic changes include change in shell outline from subovate to elongate-ovate to labiatoid with increasing size, change from fine juvenile to coarser and regularly alternating adult ornamentation, and change in beak characteristics with increasing growth from moderately prosogyrous and slightly projecting to strongly prosogyrous and moderately projecting.

Shell very thin, slightly thickened at hinge. Other interior characters not observed on Oklahoma specimens, assumed to be same as for genus.

Remarks. *Mytiloides mytiloides* is distinguished from other members of the lineage in the Hartland and Bridge Creek Members of Oklahoma by a combination of characters: its elongate, highly inclined (prosocline) shell; its projecting, prosogyrous, moderately inflated beaks and umbos; its closely spaced juvenile rugae; and its regular alternation of strongly and evenly developed, subequally spaced rugae and moderately coarse raised growth lines on adult portions of the shell. Isolated occurrences of these characters are also found in some other *Mytiloides*. *M.* sp. aff. *M. submytiloides* has a roughly similar shape but is less convex, has faint, irregular concentric ornament, and a faint posteroventral sulcus. *M. opalensis* is rounder is outline, less convex, and lacks raised even growth lines except in restricted areas of some specimens. *Mytiloides* sp. transitional between *M. opalensis* and *M. mytiloides* is a more rounded, less inclined shell with similar but finer concentric ornament and a projecting anterior margin just below the beak.

Mytiloides labiatus (Schlotheim), s. l., a species to which this form has been assigned for years, is a possibly contemporaneous or younger form that Seitz (1934 [1935], Figs. 9, 11, Pl. 38, fig. 1) has clearly shown to have much more irregular, more closely spaced rugae and flattened growth lines, a much more inflated, erect, projecting beak and umbo, and a higher angle of inclination than typical *M. mytiloides*.

Mytiloides mytiloides s. s. gives rise to two distinct groups of species in the Western Interior: one has a rounded, flat shell with equally alternating concentric rugae and raised growth lines—the group of *M.* "*latus*" (in the sense of Hattin, 1962); the second has a strongly prosocline shell with much finer concentric ornament—the group of *M. labiatus* s. s. and *M. hercynicus*. Both are descendants of *M. mytiloides*. *M. latus* and *M. hercynicus* are found in the uppermost Bridge Creek Limestone Member of the Greenhorn Formation and lower Fairport Member of the Carlile Shale elsewhere in the Western Interior.

Occurrence. *Mytiloides mytiloides* (Mantell) occurs throughout the upper one-third of the exposed part of the Bridge Creek Limestone Member (Early Turonian), Greenhorn Formation, in units 96, 98, and 100–102, section 2. The specimens occur primarily as relatively complete single valves in light gray, hard, resistant, sublithographic to very fine grained, thin ledge-forming limestones. *Mytiloides mytiloides* is the key zone fossil of this interval throughout the Western Interior and in many other areas of the world and is probably the most cosmopolitan of the Inoceramidae.

Types. Illustrated hypotypes from unit 100, section 2: internal or composite molds of adult left valves, USNM 169383 Pl. 6, fig. 16), USNM 169385 (Pl. 6, fig. 13); adult right valve, USNM 169384 (Pl. 6, fig. 14); two molds of juvenile left valves, and latex casts, USNM 169386a, 169386b (Pl 6, figs. 12, 15); a juvenile right valve, USNM 169387 (Pl. 6, fig. 11). Unillustrated, but measured or described specimens from section 2: USNM 169389 (unit 96), USNM 169390 (unit 100), USNM 169391 (unit 102).

Mytiloides sp. transitional between *M. opalensis* (Böse) and *M. mytiloides* (Mantell) (Pl. 6, fig. 4; Pl. 7, fig. 3)

Material. Three nearly complete composite molds of adult valves, two right, one left, with distal portions of ventral or posterior flanks broken off; a few additional fragments.

Description. Maximum observed height 40 mm, length 28 mm, width 6.6 mm; incomplete specimens indicate heights to 60 mm. Outline (Pl. 6, fig. 4) diagonally ovate, moderately prosocline, angle of inclination averages 69°. Anterior edge gently, convexly curved with slightly greater curvature just below beaks; ventral margin moderately and asymmetrically curved; posterior marginal curvature convex, slight to moderate; posterodorsal hinge margin straight, its length equal to 0.6 of total valve length (Pl. 6, fig. 4). Slightly biconvex; left valve slightly more inflated; maximum inflation dorsocentral along growth axis at base of umbo. Anterior flank moderately steep; ventral and posterior flanks gently inclined. Posterior auricle subtriangular, with curved posterior flank; flattened, subtly differentiated from main body of disc by postumbonal break in slope, forming shallow sulcus. Beaks, umbos, subequal, left one very slightly larger and more inflated than right; blunt, narrowly rounded, slightly projecting above hinge line on both valves (Pl. 6, fig. 4).

Juvenile, adult ornament similar, consisting of weakly to moderately and subequally developed, subequally (stratigraphically younger forms) to irregularly spaced (stratigraphically older forms) rugae gradually becoming more distantly spaced with growth; rugae small on umbo, moderately developed at midshell, becoming weaker and distantly spaced on older portions of shell. Fine, equally spaced and evenly developed, crowded raised growth lines lie between and on rugae over entire shell surface; 5 to 10 growth lines per interspace (Pl. 6, fig. 4). Earliest forms (units 86 and 88, sec. 2) have more widely and irregularly spaced rugae and finer, more numerous, closer growth lines than younger forms (unit 96, sec. 2), which have regular coarser rugae and growth lines. Interior morphology and shell unknown.

Remarks. This form is mostly closely related to, and has the general shape of, *Mytiloides opalensis*. It differs from that species in having regular raised growth lines over the entire shell, less prominent rugae, and much finer juvenile ornament and in being less rounded, without a prominent earlike projection of the anterior margin below the beaks. The ornament, though finer, resembles that of *M. mytiloides*—the more inclined, more convex, younger species of the lineage that is derived from the transitional form. Early representatives of this transitional form have a more irregular, fainter ornament and are somewhat more inclined—characters that seem to link it evolutionarily with older *M.* sp. aff. *M. submytiloides* (Seitz). This transitional form, possibly a distinct species, is thus an important evolutionary link between older and younger members of the lineage. Better material is necessary to document fully its morphology and systematic relationships.

Geographic and Stratigraphic Position. From units 86, 88, 90, and 96, section 2, in the middle part of the Bridge Creek Limestone Member, Greenhorn Formation, local zone of *M. opalensis* and basal bed of the *M. mytiloides* zone.

Figured and Described Specimens. Small adult right valve, typical shape and ornament, USNM 169311 (unit 86, sec. 2; Pl. 6, fig. 4); a large worn left valve with posterior margin missing, USNM 169313 (unit 90, sec. 2); a small worn right valve, USNM 193495 (unit 96, sec. 2).

Mytiloides opalensis (Böse)
(Pl. 6, figs. 3, 6)

Inoceramus opalensis Böse, 1923, p. 184, Pl. 13, figs. 1-3. Tsagareli, 1963, p. 79, Pl. 6, fig. 2.

?*Inoceramus* sp. aff. *I. opalensis* Böse. Adkins, 1928, p. 96.

Inoceramus labiatus var. *opalensis* Böse. Seitz, 1934 (1935), p. 457, 458, Pl. 38, figs. 4-6; Pl. 39, figs. 13-16.

Inoceramus labiatus var. *opalensis* forma *elongata* Seitz, 1934 (1935), p. 458-465, Figs. 14, 15.

Table 6. Summary of Measurements for *Mytiloides opalensis* (Böse)

Character*	USNM 169310a	USNM 169310b	Average
H	34.5 mm	19.9 mm	27.2 mm
L	37.6 mm	19 mm	28.3 mm
W	3.9 mm	4.3 mm	4.1 mm
LHL	9.5 mm	8 mm	8.75 mm
a	119°	111°	115°
i	51°	45°	48°
R/AL_j	8	8	8
R/AL_1	12	. .	12
R/AL_2	7	. .	7

*Character measurements explained in Figure 1. R = number of rugae. AL_j = axial length of juvenile part of shell (that is, the first 15 mm). AL_1 = first 25 mm of axial length. AL_2 = second 25 mm of axial length.

Material. Two nearly complete internal molds with well-preserved ornament, right valves (1 juvenile and 1 moderate-sized adult). Some additional fragments showing principal shell features.

Description. Summary of measurements given in Table 6. Shell attaining moderate size, maximum recorded height 39 mm; subequivalve, slightly biconvex, maximum convexity dorsocentral; anterior flank moderately sloping, other flanks slightly inclined to flattened. Shell moderately prosocline (average angle of inclination 55°); beak-umbo nearly terminal, at anterior end of hinge line, suberect, slightly prosogyrous, very slightly projecting above hinge line (Pl. 6, fig. 3). Valve outline inclined ovate with evenly, moderately, and convexly curved ventral and posterior margins, slightly curved anterior margin except below beak where anterior margin projects prominently, forming earlike structure (Pl. 6, fig. 3); dorsal margin of moderate length, straight, equaling more than one-half total shell length. Posterior auricle small, subtriangular, flattened, continuous with disc, separated from it only by break in postumbonal slope rather than by distinct sulcus. Juvenile shells more rounded than adults, with relatively shorter dorsal margins.

Juvenile and adult ornamentation very similar. Juvenile ornament (to about 13 mm of length along midline axis) very coarse for the lineage (Pl. 6, fig. 6), consisting of moderately large, subequally developed and subevenly spaced, asymmetrical rugae with steep ventral slopes; interspersed growth lines lacking or very sparse and incomplete on juvenile shell (Pl. 6, fig. 6). Adult ornament consisting of similar but coarser and progressively more widely spaced, subequal and subevenly spaced rugae (Pl. 6, fig. 3), in some cases crowded near the ventral margin. Most specimens lack raised growth lines or lamellae altogether; a few have fine raised incomplete growth lines between and on rugae over certain parts of shell, especially on the middle of the disc at later growth stages (Seitz, 1934 [1935], Fig. 14c; Böse, 1923, Pl. 13, figs. 2, 3); raised growth lines crowded and subregular. All gradations of ornament noted.

Shell very thin, slightly thickened at hinge line. Pallial line thin, weak, apparently entire, near margin, incompletely known. No other internal features preserved on Oklahoma specimens.

Remarks. *Mytiloides opalensis* (Böse) is distinct from *M.* sp. aff. *M. submytiloides* (Seitz) in that *M. opalensis* has a rounded form, more regular adult ornament, and coarser juvenile ornament and lacks a sulcus between the posterior auricle and umbonal fold of the shell. It is distinct from the closely related and somewhat younger *M. mytiloides* in being rounder, more erect, and less convex and in having very slightly projecting suberect beaks (as compared to strongly prosogyrous projecting beaks in *M. mytiloides*), a projecting dorsoanterior margin, coarse juvenile rugae not greatly different from that of the adult, and somewhat less regular, more closely spaced adult rugae normally lacking raised growth lines between them. *Mytiloides opalensis* is most similar to the form here identified as *M.* sp. transitional between *M. opalensis* and *M. mytiloides.* From the transitional form, *M. opalensis* differs primarily in lacking regular raised growth lines over the entire shell and in having coarser rugae, especially during early growth stages. The two are roughly gradational through forms of *M. opalensis* that have traces of fine raised growth lines.

The species also resembles *M.* "*latus*" Mantell (in the sense of Woods, 1911, p. 284, 285, Figs. 39-41 and Hattin, 1962) in general characteristics, but differs in being more ovate (*M.* "*latus*" is typically subrounded), slightly more inclined, with a shorter dorsal margin and a smaller posterior auricle. *M. latus* has finer juvenile ornamentation and typically more regular adult alternation of rugae and complete growth lines over the entire shell.

Seitz (1934 [1935], p. 457-465) divided Böse's species into two forms, an erect form typical of the primary types (Böse, 1923, Pl. 13, figs. 1-3) and a more inclined, elongate form (forma *elongata* Seitz). Both forms have similar early adult and juvenile growth stages (the size range represented by Oklahoma specimens) and are separable primarily on the basis of older adult features. The small Oklahoma specimens are not divisible into described subspecies, varieties, or forms.

Stratigraphic and Geographic Occurrence. *Mytiloides opalensis* occurs in the middle one-third of the Bridge Creek Limestone Member, Greenhorn Formation, units 88 and 90, section 2. It lies between the zones of *Mytiloides* sp. aff. *M. submytiloides* and *M. mytiloides* (Mantell). Specimens were found only in light gray, cream- and buff-colored, very fine grained, dense, burrow-mottled limestone in this interval.

Types. Hypotypes. A block containing four valves with weak growth lines and coarse rugae on one bedding plane: USNM 169310a, 169310b, 169310c, 169310d (unit 90, sec. 2; Pl. 6, figs. 3, 6). A left valve transitional between forms with and without regularly intercalated growth lines, USNM 169312 (unit 88, sec. 2).

Mytiloides sp. aff. *M. duplicostatus* (Anderson)
(Pl. 6, fig. 5; Pl. 7, figs. 2, 6)

Refer to *Inoceramus duplicostatus* Anderson, 1958, p. 100, Pl. 17, figs. 3, 4.

Material. Three specimens, one a large fragment of a left valve, the others relatively complete composite molds of a left and a right valve, both with the umbonal crest and portions of the posterior flank broken off.

Description. Moderately small for lineage, attaining height of approximately 65 mm, length of 45 mm, and width of 11 mm (largest specimen). Outline elongate-ovate (Pl. 6, fig. 5), moderately prosocline, angle of inclination averaging 54°. Anterior, posterior margins gently and convexly curved; slight projection of anterior margin just below beak, no auricle formed. Ventral margin moderately, asymmetrically curved; straight posterodorsal margin (hinge line) equal to about one-half of total shell length. Moderately biconvex with left valve slightly larger and more inflated. Maximum inflation dorsocentral at ventral edge of umbo; anterior flank steep; ventral and posterior flanks gently sloping. Beak blunt to narrowly rounded, slightly projecting above hinge line (Pl. 6, fig. 5). Posterior auricle flattened, subtriangular, indistinctly separated from disc by moderate change in posterior slope angle; no sulci or umbonal fold developed. Adult ornament consisting of moderately raised, asymmetrical, regularly to subregularly developed rugae (8 in 25 mm of height at midshell); rugae commonly paired or with two closely spaced, coarse raised growth lines on their crests, which produce a double ridge (Pl. 7, figs. 2, 6). One to three moderately coarse raised growth lines regularly to irregularly developed in interspaces between rugae and parallel to them; raised growth lines becoming fainter (Pl. 6, fig. 5) and partially disappearing (Pl. 7, fig. 2) on adult portions of shell, in some cases replaced by scattered flat lamellae. Hinge characters and musculature unknown. Only prismatic layer of shell preserved; very thin (0.25 to 0.5 mm) over most of disc, slightly thickened around ligamental area.

Remarks. The coarse, double-ridged rugae, and the moderately coarse raised growth lines that diminish in size or disappear with age serve to distinguish this species easily from all others in the lineage that are found in Oklahoma. *Mytiloides duplicostatus* Anderson from Lower to Middle Turonian strata of the Pacific coast is a much more inclined form, with a more projecting beak and umbo, a relatively longer, more curved shell, and growth lines that are initially finer and become coarser and more prominent with age (Anderson, 1958, Pl. 17, figs. 3, 4). It is specifically distinct but related to the Oklahoma form. Formal naming and detailed description of the Western Interior species await better material, however.

Compared with other *Mytiloides* in Oklahoma, *M.* sp. aff. *M. duplicostatus* is closest in shape to *M. opalensis* (Böse) and forms the transition between *M. opalensis* and *M. mytiloides* (Mantell) (see Pl. 6, figs. 3, 4). *M.* sp. aff. *M. duplicostatus* differs from both in having double-ridged rugae, coarse growth lines diminishing with age, irregular concentric ornament, and a less projecting anterior margin below the beaks. *M.* sp. aff. *M. duplicostatus* lacks the regularly alternating ornament and more inclined, curved shell of *M. mytiloides.*

The Oklahoma occurrence, from a single limestone bed, probably does not represent the full biostratigraphic range of this species. But its co-occurrence with *M. opalensis* and the lack of even fragments of it in younger or older zones probably indicate that it is primarily restricted to the *M. opalensis* zone. *M.* sp. aff. *M. duplicostatus* is probably an evolutionary derivative of early *M. opalensis.*

Stratigraphic and Geographic Occurrence. All known specimens are from unit 90, section 2, lower-middle part of Bridge Creek Limestone Member, Greenhorn Formation, in dense, very fine grained, gray argillaceous limestone.

Figured and Described Specimens. A relatively complete left valve, USNM 169313; (Pl. 6, fig. 5); partial right valve, USNM 169388 (Pl. 7, fig. 2); large fragment with typical ornament, USNM 169392 (Pl. 7, fig. 6).

Mytiloides sp. cf. *M. submytiloides* (Seitz)
(Pl. 6, figs. 7, 10)

Refer to *Inoceramus labiatus* n. var. *submytiloides* Seitz, 1934 (1935), p. 444, Fig. 6-8, Pl. 37, figs. 1-3.

Material. Two nearly complete internal molds of adult right and left valves; umbos poorly preserved, valve margins broken extensively.

Description. Valves attaining length of 45 mm; see Table 7 for summary of measurements. Subequivalve, slightly biconvex with maximum inflation dorsocentral near junction of juvenile and adult ornamentation; ventral flank slightly sloping, anterior flank moderately sloping, posterior flank and auricle flattened. Moderately prosocline, average angle of inclination 42°. Valve outline elongate-ovate, labiate, posteroventrally extended (Pl. 6, fig. 10); curvature of anterior and dorsoposterior margins slightly convex; ventral and posteroventral margins moderately and unevenly convex. Hinge line relatively short, straight, only slightly thickened in ligamental area. Beaks blunt; beak-umbo terminal-anterior, moderately prosogyrous, very slightly projecting above hinge line. Umbonal fold broad, asymmetrical, steepest posteriorly, separated from posterior auricle by shallow, posteroventrally directed sulcus (Pl. 6, fig. 10). Posterior auricle (Pl. 6, fig. 10) poorly defined, flattened, subtriangular, of moderate size for lineage.

Ornament of umbo (juvenile shell) to 13 mm of shell height consisting of small, subequally developed, subevenly spaced, faint low rugae with very fine, flat lamellate growth lines between and on them where preserved (preservation poor on both specimens). Adult ornament irregular, weakly developed, consisting of both low, subregular, moderately spaced rugae, and irregularly developed, normally widely spaced coarser rugae (Pl. 6, figs. 7, 10). All rugae of small to moderate size for genus, but very low, broad, asymmetrically rounded. Fine, slightly raised, crowded growth lines of highly variable strength cover shell, rarely alternating regularly with rugae as in typical *M. mytiloides* (compare fig. 10 with fig. 16 on Pl. 6). Shell very thin, slightly thicker near hinge line.

Ontogenetic changes include change from subequally developed to irregular ornament and from subquadrate to elongate-ovate shell form.

Interior structures of hinge and beak area not preserved. Internal mold of one specimen marked with faint raised radial lines on dorsal half—the impressions of mantle (pallial) rectractor muscle tracks. Other muscle insertion areas not preserved.

Remarks. This species is distinguished from other described Cenomanian-Turonian *Mytiloides* by its irregular, weakly developed concentric ornamentation and its shallow posteroventrally trending sulcus. Most members of the *M. labiatus* lineage are nonsulcate and have regular to subregularly developed surface ornament consisting of either alternating rugae and raised growth lines (Pl. 6, fig. 16) or only regular rugae (Pl. 6, fig. 3). Compared to typical "*I. labiatus* n. var *submytiloides*" (Seitz, 1934 [1935], Fig. 8, Pl. 37, fig. 1), the poorly preserved Oklahoma specimens are somewhat broader and less inclined, have a slightly better developed auricular sulcus, have scattered coarser rugae, and lack the anterior sulcus and truncated anterior margin shown by some of Seitz's types. Preservation of the Oklahoma specimens is not good enough to determine whether these differences are sufficient to propose a distinct subspecies, however.

It has long been assumed that *M. labiatus* (Schlotheim) (now *M. mytiloides* (Mantell)) was the earliest member of the lineage to appear in the Western Interior during Cretaceous time, that its appearance was abrupt, marking the base of the Turonian Stage, and that it was the only species of *Mytiloides* found in lowest Turonian strata. Kauffman's major revision of American *Mytiloides*, now under way, has shown these assumptions to be false.

Table 7. Summary of Measurements for *Mytiloides* sp. aff. *M. submytiloides* (Seitz)

Character*	USNM 169334	USNM 169335	Average
H	47 mm	34 mm	40.5 mm
L	42 mm	32 mm	37 mm
W	6 mm	5 mm	5.5 mm
LHL	. .	6 mm	. .
a	86°	89°	87.5°
i	43°	41°	42°
R/AL_j	. .	13	. .
R/AL_1	. .	18	. .
R/AL_2	7	. .	. .

*Character measurements explained in Figure 1. R = rugae. AL_j = axial length of juvenile part of shell (that is, the first 15 mm). AL_1 = first 25 mm of axial length. AL_2 = second 25 mm of axial length.

Intensive collecting and population analysis have demonstrated the following evolution for the *M. labiatus* lineage throughout the Western Interior: (1) The lineage began as a series of rare, small- to moderate-sized, weakly ornamented forms (such as *Mytiloides* sp. aff. *M. submytiloides*) in uppermost Cenomanian rocks dominated by *Inoceramus pictus* and related species. These forms ranged through the lower part of the Bridge Creek Limestone member (zone of *Sciponoceras gracile*) of Late Cenomanian age and into basalmost Turonian strata. (2) Forms normally ascribed to "*M. labiatus*" in the Lower Turonian part of the Bridge Creek Limestone Member and associated rocks are divisible into four and locally five distinct species and subspecies, which thus allows finer biostratigraphic division of the member. (3) The *M. labiatus* lineage continued into Coniacian rocks without a break and gave rise to numerous biostratigraphically useful taxa.

Stratigraphic and Geographic Occurrence. *Mytiloides* sp. aff. *M. submytiloides* (Seitz) is well represented from a single level (unit 62, sec 2) in the upper part of the Hartland Member, Greenhorn Formation, regional zone of *Dunveganoceras albertense*, and is represented by poorly preserved specimens in the uppermost Hartland Member (unit 72, sec. 2) and in the lower one-third of the Bridge Creek Limestone Member (units 78 and 81, sec. 2), zone of *Sciponoceras gracile*, Greenhorn Formation. The species occurs in moderately thin beds of resistant buff chalk and chalky limestone.

Figured and Mentioned Specimens. Illustrated specimens from unit 62, section 2: an adult left valve, nearly complete, USNM 169334 (Pl. 6, fig. 7); a small adult right valve with broken margins but showing ornament, USNM 169335 (Pl. 6, fig. 10). Poorly preserved specimens from section 2: USNM 169336 (unit 72), USNM 169337 (unit 78), USNM 193499 (unit 82).

Mytiloides? *crippsi* Lineage
Mytiloides? sp. aff. *M.*? *crippsi* (Mantell)
(Pl. 5, fig. 1)

Refer to *Inoceramus crippsi* Mantell. For synonymy, see Woods, 1911, p. 273–278, Figs. 34, 35.

Material. A single internal mold, dorsal half of a small adult right valve, and latex cast.

Description. Moderate size for genus, specimen probably attaining height and length of approximately 75 and 65 mm, respectively. Measured width of laterally crushed valve 4 mm. Prosocline; angle of inclination 63°; more erect than typical *M.*? *crippsi*; somewhat prosogyrous; beak blunt?, anterior, terminal; umbo flattened, differentiated from main disc by shallow ruga and change in ornament (Pl. 5, fig. 1); beak, umbo probably very slightly projecting above hinge. Maximum convexity near base of umbo; ventral, posterior flanks flattened, anterior flank somewhat inclined on crushed valve. Outline subcircular to broadly subovate (Pl. 5, fig. 1; Woods, 1911, Fig. 35); growth line trace indicates broadly rounded (convex outward) anterior and posterior flanks, moderately and subevenly rounded ventral flanks. Dorsal margin long, straight (Pl. 5, fig. 1). Posterior auricle moderately large, subtriangular with rounded apex, subtly defined by slight break in dorsoposterior slope and possibly shallow auricular sulcus; posterior auricular commissure slightly folded on Oklahoma specimen. External ornament reflected on shell interior (Pl. 5, fig. 1) consisting of fine, faint, equally developed and evenly spaced, slightly raised growth lines or lamellae on umbo (juvenile shell). Similar but less regular and more widely spaced growth lines are superimposed on faint, narrow to broad, irregularly and widely spaced rugae on preserved dorsal half of adult shell.

Ligamental plate abruptly thickened with a prominent fold of prismatic layer; layer otherwise very thin; no nacre preserved. Subumbonal cavity seemingly narrow and of moderate depth, although shell is compressed laterally. Umbonal septum small, unnotched, a slightly thinner diagonal extension of the ligamental plate to the anterior margin, which is dorsally somewhat thickened and bears near its upper edge a small narrow projection or pseudotooth on this specimen (Pl. 5, fig. 1). Resilifers moderately broad, 5.5 in 10 mm, separated by narrowly rounded ridges; resilifers facing upward and slightly inward, as shown in *M.? crippsi* by Woods (1911, Pl. 48, fig. 2b). Faint narrow internal rib discontinuously extending from hinge posteroventrally along junction of auricle and disc. Dorsoposterior pallial line entire, prominent, a series of narrow fine parallel ridges; other musculature not preserved.

Remarks. The generic assignment of "*Inoceramus*" *crippsi* Mantell is uncertain at present. Most authors put it with *Inoceramus*, as apparently did Cox (1969, p. N315) by including similar forms. Donovan (1954, p. 17) included it in *Inoceramus* (*Inoceramus*). Subsequent studies of interior morphology and hinge characteristics by Kauffman and review of exterior features in Inoceramidae from North America and Europe have shown that *Inoceramus cuvierii* Sowerby, the accepted genotype (Cox, 1969, p. N315) is grossly distinct from "*I.*" *crippsi* in being strongly inequivalve, in hinge characters, ornament, musculature, nature of the byssal slit area, and in the character of the posterior auricle and strong auricular sulcus. The two are different enough to warrant generic separation. Furthermore, the subgenus *Inoceramus* (*Inoceramus*) is illustrated (Cox, 1969, Figs. C46, 1a, 1b, 1c) by *I.* (*I.*) *cuvierii* and *I.* (*I.*) *lamarcki* Parkinson; the latter, a true *Inoceramus* (*Inoceramus*) is even more grossly (generically) distinct from "*I.*" *crippsi* than *I.* (*I.*) *cuvierii*. Thus, neither Cox nor Donovan correctly assigned "*I.*" *crippsi* and related forms. Among other genera accepted as valid by Cox (1969, p. N315–N321), *I.* (*Mytiloceramus*) of Middle Jurassic age closely resembles "*I.*" *crippsi* in external form and relative valve size but has quite different musculature and a small anterodorsal auricle and is generically distinct.

Although the form of "*I.*" *crippsi* is distinct from typical *Mytiloides* (type, *M. labiatus* Schlotheim), the latter genus, as redefined elsewhere in this paper, also includes more rounded forms descended directly from or ancestral to mytiliform species (for example, the *M. labiatus* to *M.* "*latus*" series). The "*I.*" *crippsi* lineage shows similar though less pronounced variation (see Woods, 1911, series from "*I.*" *crippsi* as in Pl. 48, fig. 3, to "*I.*" *crippsi reachensis* Etheridge, Pl. 48, fig. 5); in addition there are many other characters in common with *Mytiloides*. Among these are subequivalve, thin, slightly to moderately convex shells; alternating subregular rugae and raised growth lines or lamellae over the valves; terminal, rounded beaks and umbos that are not greatly projecting; and similar musculature (entire pallial line, laterally elongate posterior adductor, no visible pedalbyssal insertion areas). The "*I.*" *crippsi* and *M. labiatus* lineages differ mainly in the thick ligamental plates, open centrodorsally facing ligamental areas, and much coarser resilifers of "*I.*" *crippsi*. "*I.*" *crippsi* also retains an internal rib situated between the posterior auricle and the disc, a feature present on many older, related species, reflecting its earlier Cretaceous ancestry. This structure is lost in the *M. labiatus* lineage.

"*Inoceramus*" *crippsi* is therefore not an *Inoceramus* s. s. and shows lesser but significant differences from *Mytiloides*. Pending a modern generic revision of the Inoceramidae based on internal and external features, "*Inoceramus*" *crippsi* and related species are assigned provisionally to the most similar recognized Cretaceous genus, *Mytiloides*; the "*I.*" *crippsi* lineage very likely represents a distinct genus or subgenus ancestral to true *Mytiloides*.

The single Oklahoma specimen shows only the dorsal part of the shell interior well. The heavy, dorsocentrally facing ligamental plate, faint internal rib, coarse resilifers, and nearly flat, subrounded shell clearly place it in the *M.*? *crippsi* lineage rather than other Early and early Late Cretaceous lineages. Among described forms in this lineage, the very faint, fine juvenile ornamentation, subtle and irregular nature of the rugae, shell form, and very reduced beak and umbo most closely resemble rounded specimens of *M.*? *crippsi* (Mantell), s. l., such as that illustrated by Woods (1911, Pl. 48, figs. 2a, b) from the Upper Greensand zone of *Pecten asper*, probably the Lower Cenomanian Substage as currently defined in Europe. *M.*? *crippsi reachensis* Etheridge has a similar form but has regularly alternating rugae and growth lines, not found in the Oklahoma specimen. "*Inoceramus*" *comancheanus* Cragin and "*I.*" *bellvuensis* Reeside represent a group (possibly the same genus) from which this lineage could have been derived; both have even, closely spaced, small rugae over the entire shell and strong internal ribs between the auricle and the disc.

Stratigraphic and Geographic Position. Upper shale member of the Graneros Shale in soft, thin and evenly laminated, dark olive gray, noncalcareous clay shale (unit 10, sec. 2).

Illustrated Specimen. Internal mold and latex cast, dorsal half of moderate-sized adult right valve, USNM 172868a, 172868b.

Superfamily PECTINACEA Rafinesque, 1815
Family ENTOLIIDAE Korobkov, 1960

Genus *Entolium* Meek, 1865
Subgenus (*Entolium*): Emended

Diagnosis. Shells small, subcircular, subovate, or suborbicular, slightly biconvex, very nearly equivalve, subequilateral, orthogyrous. Shell very delicate, thin, composed of inner calcite foliate layer, outer radially fibrous layer in both valves. Auricles subequal, small, without byssal notch or opening at any growth stage, commonly projecting above hinge as angular wings on left valve, but ranging to forms with nonprojecting auricles whose dorsal margins slope slightly downward, away from beak (as in *E.* (*E.*) *gregarium*, n. sp.). Lateral commissures closed. Ornament of very fine, faint, closely spaced, regularly to irregularly developed growth lines; or shell smooth. Hinge area with well-defined auricular crura, and with central triangular resilifer having narrow, incised, linear, lateral extensions on auricles.

Remarks. The shape, ornament, shell structure, and hinge characters of Cretaceous species from the Lincoln(?) and Hartland Members of the Greenhorn Formation in the central Western Interior define this as *Entolium* (*Entolium*), but the small auricles do not project above the hinge line on either valve and in many cases have their dorsal margins sloping gently downward from the beak. Furthermore, the auricles are subequal in size on left and right valves. These characteristics of the auricles distinguish the American Cretaceous form from typical *Entolium*

Table 8. Summary of Measurements for *Entolium* (*Entolium*) *gregarium*, n. sp.

Character*	No. of specimens	Range	Average
H	22	4–9 mm	6.9 mm
L	22	3.5–8.5 mm	6.5 mm
LHL	22	1–3.5 mm	2.4 mm

*Character measurements explained in Figure 1.

(*Entolium*) as defined previously (Newell and Hertlein, 1969, p. N346, N347). We do not consider these differences as being of generic or subgeneric importance, as they involve only the shape of the auricles in the left valve, and thus we have expanded the concept of the subgenus to include them.

Entolium (Entolium) gregarium Kauffman and Powell, n. sp.
(Pl. 4, fig. 2; Pl. 5, fig. 4; Pl. 6, figs. 8, 9; Pl. 8, figs. 1, 2)

Material. Approximately one hundred small complete valves, many with shell and interior features preserved but laterally crushed; crowded on single bedding planes.

Description. Shell small, attaining a maximum height of 9 mm; subequivalve, subequilateral, very slightly prosocline, very slightly biconvex, with maximum inflation dorsocentral near ventral umbo; all flanks and auricles flattened. Orthogyrous, beaks small, narrowly rounded, central, depressed to very slightly projecting above dorsal margin. Auricles subequal on both valves and between valves, subtriangular (Pl. 8, fig. 2), small, their dorsal margins straight, subparallel to hinge line, or slightly sloping downward away from beak; projecting corner of auricles somewhat rounded. Auricular sulci absent or very faint; broad, shallow. Outline subround (Pl. 8, figs. 1, 2), with commissure evenly and moderately curved except below auricles, where straight to very slightly concave. Surface covered with very fine, faint, closely and subequally spaced growth lines. No radial ornament. Preserved shell very thin, white in color, calcitic.

Muscle insertion areas unknown. Hinge area consists of two very narrow but prominent auricular crura, which extend at least one-half way from beak in position of auricular sulci, and two short narrow extensions of central triangular resilifer, subparallel to, but below, the dorsal auricular margin (Pl. 4, fig. 2; Pl. 8, fig. 2). Dorsal margin of auricles slightly thickened, forming a low internal ridge. Summary of measurements given in Table 8.

Remarks. This very small, delicate species has been known for some time in the Lincoln(?)-Hartland interval of New Mexico, Colorado, and Kansas, but well-preserved material has been lacking until the discovery of the Oklahoma specimens. The latter are also the largest yet found and show nearly all interior and exterior features, which permit description and confident generic assignment.

Although isolated specimens are known, the species occurs dominantly in dense accumulations of hundreds of specimens on single bedding planes in the shale. Double and single valves are both common, and most specimens are complete and well preserved. A semigregarious life habit similar to many modern deep-water pectinids is indicated, and the excellent preservation of these bottom clusters, probably in situ, is accredited to mass mortality due to either the rapid deposition of a thin mud cover (possibly a microturbidite flow) or the local development of lethal chemical and (or) low-oxygen situations near the water-substrate interface.

Stratigraphic and Geographic Occurrence. *Entolium* (*Entolium*) *gregarium*, n. sp., occurs throughout the lower and middle parts of the Hartland Member, Greenhorn Formation, regional zone of *Dunveganoceras albertense*, in units 1, 2–4, 5, 9, 13, 15, 22, and 24, section 3. The specimens of units 1, 5, and 9 are field records only. Two poorly preserved "paper pectens" that represent this lineage occur in unit 63, section 2 (lower part of upper Hartland). They are listed as *Entolium*? sp. aff. *E. gregarium*.

Types. Holotype, complete composite mold of an adult valve showing hinge characters, USNM 169338a (unit 3, sec. 3; Pl. 8, fig. 2). Paratype on same slab, incomplete adult valve showing hinge characters, USNM 169338b (Pl. 4, fig. 2); other paratypes on same slab, complete small adult valves typical of species, 169338c,

169338d, 169338e, 169338f, and 169338g; a slab of small adult valves, paratypes, packed on a single bedding plane of unit 3, section 3, USNM 169339, two of which are illustrated (Pl. 8, fig. 1), USNM 169339a (left) and 169339b (right); a paratype slab with several adult individuals on a single bedding plane of unit 3, section 3, USNM 169340, with illustrated specimens designated USNM 169340a, 169340b (Pl. 6, fig. 8, left and right, respectively), and 169340c (Pl. 6, fig. 9). Unfigured paratypes, single slabs with abundant specimens. USNM 169341 (unit 3, sec. 3), USNM 169343 (unit 4, sec.3), USNM 169342 (unit 13, sec. 3), USNM 193500 (unit 15, sec. 3), USNM 169344 (unit 22, sec. 3), and USNM 169345 (unit 24, sec. 3); a questionably related specimen, USNM 169346 (unit 63, sec. 2).

Genus *Syncyclonema* Meek, 1864

Generic references in Speden, 1967; Newell and Hertlein, 1969, p. N347.

Syncyclonema? sp.

Material. A single poorly preserved internal mold, incomplete dorsally, of a small adult(?) valve.

Remarks. The poorly preserved specimen from the lower part of the Hartland Member of Oklahoma probably represents an undescribed species of *Syncyclonema* common elsewhere in the slightly younger *Sciponoceras gracile* Zone of the Bridge Creek Member, Greenhorn Formation. Speden (1967) gave the general characteristics of Cretaceous *Syncyclonema.* The Bridge Creek specimens are small, ovate in outline, nearly acline and equilateral, being slightly inclined anteriorly. The commissural margins are evenly and moderately rounded except below the auricles, where they are slightly concave. Both auricles are small, triangular, and flat, the anterior one larger and anteriorly extended along its dorsal margin. The shell is slightly biconvex, with maximum inflation dorsocentrally; the beaks are small, narrowly rounded, centrally situated, and do not project significantly above the hinge line. The ornament, known only from impressions on the internal mold, consists of fine, weakly developed, closely spaced growth lines. The largest known shells reach only 21 mm in height. Formal description of this species should await collections having the outer shell surface and auricular characters well preserved. Compared to described Cretaceous species from the Western Interior, the Greenhorn species is more rounded and less coarsely ornate than the Campanian *S. halli* (Gabb), the most closely related form.

Stratigraphic and Geographic Occurrence. Regional zone of *Dunveganoceras albertense*, lower part of Hartland Member, Greenhorn Formation, in grayish-orange, slightly resistant, moderately thin bedded chalk (unit 4, sec. 3).

Described Specimen. USNM 172883.

Suborder OSTREINA
Superfamily OSTREACEA
Family GRYPHAEIDAE
Subfamily PYCNODONTEINAE

Genus *Pycnodonte* Fischer de Waldheim, 1835
Subgenus *Phygraea*? Vyalov

Remarks. In the *Treatise on Invertebrate Paleontology*, Stenzel (1971, p. N1105–1114) revised the generic systematics of ostreids and distinguished seven genera and subgenera among the Pycnodonteinae. Gryphioid pycnodontids such as *P. newberryi* and *P. aucella* seem to have characters most similar to those of *Pycnodonte* (*Phygraea*) as defined by Stenzel: a highly convex left valve, especially in the umbonal area; lack of well-defined anterior and posterior auricles along the hinge axis; a prosocline shell; a geniculate, concave posterodorsal margin; and a nearly smooth to weakly lamellate valve surface lacking radial ornament on the left valve. *Pycnodonte newberryi* differs from typical *P.* (*Phygraea*) in having well-defined vermiculate chomata (Pl. 9, fig. 5, especially left side like those of *P.* (*Pycnodonte*). Stenzel (1971) considered this an important character for differentiation of these subgenera; thus, assignment to *P.* (*Phygraea*) is here questioned. *Pycnodonte newberryi* may represent an ancestral group of oysters that gave rise to *P.* (*Pycnodonte*) and *P.* (*Phygraea*) and combined characters of both.

The lineage of gryphioid pycnodontids to which *P.* (*Phygraea*?) *newberryi* belongs is characterized by small subrounded to prosocline, ovate shells with a highly inflated left valve having a small attachment scar; the right valve is flat to slightly concave. Both shells are thick, commonly with alternating layers of vesicular and lamellar calcite. Beaks are very small, subcentral (right valve) to anterior (left valve), and not greatly projecting. The shell surface is smooth except for fine radial incised lines on the dorsal part of the right valve and low, flat, concentric growth lamellae or squamae on both valves. Commonly, a prominent sulcus separates a low posterior auricle from the disc on the left valve. The ligament area is small, nearly symmetrical, bounded laterally by well-defined, close-set sinuous chomata restricted to the dorsal part of the shell and best developed on the interior of small flat shelves that commonly bound the hinge plate outside of the commissural shelf or rim. This rim is entire and prominent. The posterior adductor muscle insertion area is round, well incised, and situated just posterior and dorsal to midshell.

Pycnodonte (*Phygraea*?) *newberryi* (Stanton)
(Pl. 8, fig. 5)

Gryphaea newberryi, Stanton, 1893 (1894), p. 60–62, Pl. 5, figs. 1–5. Shimer and Blodgett, 1908, p. 61. Grabau and Shimer, 1909, p. 472, Fig. 360. Reagan, 1927, p. 123, Figs. 10–12. Shimer and Shrock, 1955, p. 397, Pl. 155, fig. 19.
Gryphaea pitcheri Morton, White, 1876 (1877), p. 171, Pl. 17, figs. 1a–1f.
non *Gryphaea* cfr. *newberryi*, Adkins, 1928, p. 109.

Material. Two upper (right) valves, small, complete, one showing the valve interior well. Description based on these and on Stanton's and White's type specimens (USNM 8653; USNM 8775). Additional fragments and poorly preserved whole left and right valves.

Description. Summary of measurements in Table 9. Small, attaining height of 46 mm (left valve), highly inequivalve, slightly to moderately prosocline, average angle of inclination 69° (left valves), 71° (right valves). Variable in valve outline and inflation.

Left valve subovate to subtriangular and posteroventrally extended (Stanton, 1893 [1894], Pl. 5, figs. 1–5), with slightly concave posterodorsal and dorsoposterior margin, moderately narrowly rounded (convex) midventral margin at tip of posterior auricle, moderately and unevenly rounded ventral margin with slight sinus at intersection with auricular sulcus, and slightly (typical) to moderately curved anterior margin. Moderately to highly inflated, maximum inflation central to dorsocentral, below umbo; anterior and umbonal flanks steeply sloping, ventral and posterior

Table 9. Summary of Measurements for *Pycnodonte newberryi* (Stanton)

Valve	*H*	*L*	*W*	*i*
	Gryphaea pitcheri of White (1877); USNM 8653			
Left	29 mm	30.5 mm	18 mm	71°
Left	23 mm	24 mm	14 mm	62°
Right	21 mm	21.5 mm	3.5 mm	78°
	Pycnodonte newberryi Stanton (1893); USNM 8775			
Left	46 mm	37 mm	25 mm	73°
Right	33 mm	33 mm	5 mm	71°
	Pycnodonte sp. from Oklahoma; USNM 169407			
Right	11.5 mm	14 mm	2 mm	65°
		Average		
Left	32.7 mm	30.5 mm	19 mm	68.7°
Right	21.8 mm	22.8 mm	3.5 mm	73.3°

Note: Character measurements explained in Figure 1.

flanks, including auricle, moderately sloping. Axial or umbonal fold prominent, moderately arched, asymmetrical, bounded on steep posterior side by a prominent, moderately deep, asymmetrical sulcus separating posterior auricle from disc. Auricle variously developed, shaped as a narrow curved triangle, large and posteriorly strongly projecting (typical) to reduced (Stanton, 1893 [1894], Pl. 5, figs. 4, 5) and slightly projecting; gently arched. Beaks blunt, obscure; umbos very swollen with steep flanks, erect to slightly opisthogyrous, gently curved, strongly projecting above hinge line and incurved. Shell thick. Ligament area small, subsymmetrical, with small triangular shallow resilifer bounded by larger, flattened, lateral ligamental areas. Closely set, sinuous, narrow raised chomata weakly developed on inner shell margin just lateral to ligamental area. Posterior adductor insertion area subround, posterocentral in position, well incised. Variation in shape and inflation considerable in adults, especially in regard to development of posterior auricle (see Stanton, 1893 [1894], Pl. 5, figs. 1-5, for example). Ontogenetic variation trends from subround, to subovate, to posteriorly elongated subtriangular shells.

Right (upper) valve subrounded (Pl. 8, fig. 5), flat to very slightly convex or concave, maximum convexity dorsocentral. Dorsal margin, hinge line straight to slightly concave, anterior and posterior segments nearly equal, length moderate (ratio of length of the hinge line to the length of the entire shell [oriented with hinge line horizontal] is 0.64). Anterior, ventral margins moderately and subevenly rounded; posterior margin less curved on many specimens. Anterior auricle absent or poorly defined as a small flat projection of anterodorsal flank, not clearly separated from disc; posterior auricle small, subtriangular, flat, moderately well defined by marked break in postumbonal slope. Beaks blunt, rounded, very small, subcentral, very slightly projecting over hinge line; umbo flattened to slightly convex; beak-umbo orthogyrous. Surface nearly smooth, marked by very fine radiating incised lines dorsally (visible only on well-preserved specimens), and very faint, moderately spaced growth lines. Rare small shallow rugae developed ventrally on some shells. Interior of shell with prominent raised commissural rim, 1 to 2 mm inside valve margin, in position of pallial line (Pl. 8, fig. 5—internal mold with impression). Posterior adductor insertion area of moderate size, subround, moderately impressed, concave, situated just posterior and dorsal to midpoint of shell (Pl. 8, fig. 5). Long, sinuous, subparallel chomata or raised lines crowded on interior of auricles, extending ventrally from hinge line to commissural rim (Pl. 8, fig. 5); strongest development posterior. Very little ontogenetic or adult variation in form of right valve.

Remarks. *Pycnodonte* (*Phygraea*?) *newberryi* is a rare to moderately abundant element in carbonate assemblages of the central Western Interior (Kauffman, 1967a, 1970), but unusually rare in the Bridge Creek Member of Oklahoma. Although sufficient hard substrate, in the form of *Inoceramus* shells, was present on contemporaneous Bridge Creek surfaces for settlement of oyster spat, only rare ostreids are found in this mode of life, and neither specimen of *P. newberryi* was so attached. From examination of *P. newberryi* occurrences in this and other facies within the Western Interior, and from observations on the attachment "scar" of the left valve (normally very small or apparently missing), we conclude that the preferred habitat of this species was free-living or delicately attached (mainly as juveniles) directly on the sediment surface, rather than continuously on hard substrate such as large shell fragments. The inflated left (lower) valve, posterior auricle and sulcus, and small flat upper valve are adaptations to this mode of life in gryphioid oysters (Kauffman, 1969a). The scarcity of *P. newberryi* in Oklahoma then probably reflects inhospitable sediment factors, most likely a sediment which was too water saturated and soft to support pycnodontids.

Among American Cretaceous oysters, *P.* (*Phygraea*?) *newberryi* closely resembles *P.* (*P.*?) *aucella* (Roemer) from the Austin Chalk of Texas and Fort Hays Limestone Member, Niobrara Formation, of the Western Interior. It differs from *P.* (*P.*?) *aucella* in being smaller, less inclined, with a less curved midshell axis, more central beaks (especially the right valve), and in having a smaller, less extended posterior auricle and stronger auricular sulcus. The posterodorsal-dorsoposterior margin is commonly more concave in *P.* (*P.*?) *aucella*. The two species occur in, and are adapted to, very similar facies and are part of the same evolutionary lineage. Within the Bridge Creek Member and equivalent rocks of the Western Interior, specimens referred to *P.* (*Phygraea*?) *newberryi* occur at several levels and display marked evolutionary changes in convexity, valve outline, and development of the auricle and auricular sulcus. Population analysis of these will probably allow their division into distinct, stratigraphically useful species and (or) subspecies.

Occurrence. Middle part of the Bridge Creek Limestone Member, Greenhorn Formation, units 82, 86, 88 and 100, section 2 (zones of *Mytiloides opalensis* and *M. mytiloides*).

Types. Figured hypotype, a small right valve, USNM 169407 (unit 100, sec. 2); unfigured hypotypes, right and left valves, USNM 172869 (various levels, marked on specimens, sec. 2).

Subfamily EXOGRYINAE

Genus *Exogyra* Say, 1820
Exogyra sp. aff. *E. boveyensis* Bergquist

Refer to *Exogyra boveyensis* Bergquist, 1944, p. 16, 17, Pl. 4, figs. 1–13.

Material. A single small adult(?) left valve attached to a fragment of *Inoceramus* shell.

Remarks. A small undescribed species of *Exogyra* encrusting *Inoceramus* and other shell surfaces is known commonly in the Lincoln Member, Greenhorn Formation of the Western Interior; a single specimen was found in the uppermost bed of the Lincoln in Cimarron County. The shell is small, subround, with a tightly coiled opisthogryous beak and umbo. The left valve, attached almost totally to the substrate, is moderately convex with steep sides. The right valve (not known from Oklahoma), is slightly convex, ovate to subround, with the beak and umbonal coiling less pronounced. Ornament normally consists of fine growth lamellae with or (less commonly) without fine, sinuous, bifurcating costae and costellae; radial ornament is less commonly developed on the right valve. The specimen from Oklahoma is a left valve similar to this form, attached to a shell fragment, with the free valve margins broken and the preserved ornament dominantly fine lamellae without prominent costae. This species seems most closely related to *Exogyra boveyensis* Bergquist from rocks of about the same age (Coleraine Formation) in Minnesota, but in general is smaller, is more convex, and has coarser costae (where developed) and greater beak coiling than Bergquist's species.

Stratigraphic and Geographic Occurrence. Uppermost bed of the Lincoln Member, Greenhorn Formation, probably regional zone of *Dunveganoceras conditum*, in fine-grained, light brown, slabby resistant calcarenite (unit 47, sec. 2).

Described Specimen. USNM 172870.

Family OSTREIDAE
Subfamily OSTREINAE

Genus *Ostrea* Linné
Ostrea beloiti Logan
(Pl. 8, figs. 9, 10)

Ostrea beloiti Logan, 1899, p. 214, Pl. 25, figs. 7, 8. Hattin, 1965a, p. 50, Pl. 4, figs. A, B, D, G, I. Hattin, 1965b, p. 13, Fig. 3, no. 4, 9.

Material. Approximately 50 specimens, single right and left valves, whole or large fragments, original shell preserved, showing all interior and exterior features; crowded and cemented together in a thin biostrome.

Description. See Table 10 for measurements. Small for genus, attaining length of 39 mm, height of 46 mm on Oklahoma specimens (Pl. 8, fig. 10). Moderately inequivalve; left (attached) valve moderately and irregularly convex, with shape and inflation controlled mainly by attachment surface and crowding by adjacent individuals. Attachment scar of left valve flat to concave, large, up to four-fifths of total shell surface area, bearing impression of surface (oyster, *Inoceramus* shells

Table 10. Summary of Measurements for *Ostrea beloiti* Logan

	Valve	*H*	*L*	*W*	*i*
USNM 169318	Left	46 mm	39 mm	6 mm	74°
USNM 169319	Right	20 mm	7.5 mm	3 mm	83°
	Right	9 mm	6 mm	1.5 mm	80°
	Left	37 mm	28 mm	8 mm	81°
	Right	25 mm	17 mm	5 mm	78°
	Right	13 mm	11 mm	2 mm	82°
	Left	18 mm	14 mm	8 mm	79°
	Right	17 mm	15 mm	2 mm	76°
	Left	39 mm	23 mm	10 mm	83°
	Left	22 mm	15 mm	6 mm	88°
	Right	29 mm	19 mm	3 mm	71°
	Right	16 mm	11 mm	2 mm	82°
Average		24.3 mm	17.1 mm	7.6 mm* 2.6 mm†	80°

Note: Character measurements explained in Figure 1.
*Left valves.
†Right valves.

normally) to which it was attached. Distal flanks of left valve steep, rising abruptly from attachment surface. Upper (right) valve slightly and subevenly convex, maximum inflation central on disc; flanks flattened to abruptly truncated, variable. Slightly prosocline, rarely moderately inequilateral.

Marginal outline of left valves highly variable due to crowded life habit (Pl. 8, fig. 9); characteristically ovate, elongate-ovate dorsoventrally, or subtriangular (Pl. 8, fig. 10), expanding ventrally. Anterior, ventral margins moderately curved (convexly), posterior margin straight or slightly curved (convex or concave). Outline of right (upper) valves more consistent (Pl. 8, figs. 9, 10), elongate-ovate to comma-shaped with a concave dorsoposterior margin; other margins moderately rounded. Beak pointed to narrowly rounded, situated slightly posterior to midline; beak-umbo posteriorly curved to varying degrees, somewhat coiled at tip, suberect to moderately opisthogyrous, projecting well above hinge line; apical angle acute to 90°. Umbo slightly convex to flattened. Young valves rounder than adults. No auricles developed on majority of specimens. Surface sculpture consisting of fine, irregularly spaced growth lines and flat concentric lamellae, scattered broad irregular rugae, and rarely small faint plicae near shell margins.

Ligament area consisting of shallow, narrow, slightly curved subtriangular resilifer bounded laterally by broader, horizontally striated, flat, slightly convex, subequal, triangular lateral ligamental areas (Pl. 8, fig. 10). Umbonal cavity moderately deep (left valves) to shallow (right valves). A narrow raised rim, 2 to 3 mm from commissure (Pl. 8, fig. 10), partially or wholly encircles the valve interior, normally extending at least down lateral margins and marking the pallial line. Thin, elongate, slightly sinuous, subparallel, closely and equally spaced, raised chomata or "denticles" lie between raised submarginal rim and commissure over much of valve interior, characterizing species. Posterior adductor muscle insertion area subround with flattened dorsal edge, slightly but distinctly impressed, large, situated posterocentrally. Quenstedt's muscle insertion area not observed, subtle if present, possibly a small concave depression just below ventral edge of hinge plate.

Shell structure consisting of thin flat sheets of lamellar calcite, interspersed near margins with layers of crenulated lamellar calcite, reflecting chomata. Chomata suggest affinities to the Pycnodonteinae, but no distinct vesicular shell layers were noted on material studies.

Remarks. This species—easily distinguished by its size, crowded growth habit, relatively low convexity of the attached valve, and especially by the linear, closely set chomata or "denticles" around the inner shell margin—has a remarkable stratigraphic record and considerable biostratigraphic significance. It is restricted to the upper part of the Graneros Shale and basal Greenhorn Formation (Lincoln Member), scattered in shale, but mainly concentrated in one or more thin (2 to 8 cm [1 to 3 in.]), basically isochronous oyster biostromes of calcarenite that collectively have a geographic range extending from the San Juan Basin of New Mexico north to Manitoba, and from central Colorado east to Kansas and Oklahoma. Taken together, these biostromes form a marker unit that itself is only slightly diachronous, transgressing the widespread X bentonite marker bed by, at the most, a few feet of apparently continuous stratigraphic succession. Individual beds appear to be laterally continuous over areas of several hundred square miles. This widespread, continuous development of oyster beds represents an occurrence that is unknown in modern seas.

Ostrea beloiti is easily distinguished from other, similarly shaped, small encrusting Cretaceous oysters (for example, *Pycnodonte bentonensis*, *P. congesta*) by its unique chomata.

Occurrence. *Ostrea beloiti* Zone, in thin calcarenites and calcarenitic oyster coquina that weather brown to yellowish orange, units 6, 8, 10, 12-15 (units 8, 12-15 field reports only), section 2; Thatcher Limestone Member (unit 6) and lower part of the upper shale member, Graneros Shale.

Types. Hypotypes: a slab with numerous whole and fragmented shells, including the largest specimen, a left valve, from Oklahoma, USNM 169318 (unit 14, sec. 2; Pl. 7, fig. 10); a slab with numerous right and left valves, typical of the biofabric of the biostrome, USNM 169319 (unit 14, sec. 2; Pl. 7, fig. 9).

Ostrea? sp. aff. *O. anomioides* Meek

Refer to *Ostrea anomioides* Meek, in Stanton, 1893 (1894), p. 55, Pl. 1, figs. 5, 6.

Material. Several immature right (upper) valves with shell preserved, showing external morphology. No valve interior preserved.

Description. Shell small, round to broadly ovate with slight prosocline inclination and recurved suberect to slightly opisthogyrous beak and umbo. No auricles. Only right valves known; very slightly convex with maximum convexity on umbos. Beaks bluntly pointed, very slightly projecting above dorsal margin. Ornament consisting entirely of very fine, crowded growth lines and flat lamellae. Interior shell features unknown.

Remarks. The collection consists entirely of spat and juvenile shells, and of right valves with only the external morphology known. The general form of the shells suggests *Ostrea* rather than *Pycnodonte* as the genus, and none of the thin shells seems to show anything but lamellar calcite structure. Without interior features, however, the generic assignment is in doubt. The most closely related form is *Ostrea anomioides* Meek, the adult shells of which seem to have these general characters. Definite specific assignment of the Oklahoma specimens will depend upon a collection of adult left and right valves, however.

Stratigraphic and Geographic Position. Upper shale member of the Graneros Shale (unit 8, sec. 2); lower part of Lincoln Member, Greenhorn Formation, regional zone of *Dunveganoceras pondi* (units 20, 24, and 29, sec. 2); lower (units 2 and 4, sec. 3) and middle part (unit 19, sec. 3; units 58 and 62, sec. 2) of Hartland Member, Greenhorn Formation in the regional zone of *Dunveganoceras albertense*, in clay shale, buff chalky shale, and shaly limestone.

Described Specimens. USNM 172871 through 172876.

Class SCAPHOPODA Bronn, 1862
Family DENTALIIDAE Gray, 1834

Genus *Dentalium* Linné, 1758
Dentalium? sp.

A portion of an internal mold of a smooth, slender, gently curved and regularly expanding tube suggesting described Cretaceous *Dentalium* or *Laevidentalium* was found in unit 81, section 2, uppermost part of *Sciponoceras gracile* Zone, lower part of Bridge Creek Limestone Member, Greenhorn Formation. USNM 172866.

Class GASTROPODA
Superorder PROSOBRANCHIATA
Order ARCHEOGASTROPODA Thiele, 1925
Superfamily PATELLACEA Rafinesque, 1815
Family ACMAEIDAE Carpenter, 1857

Genus *Acmaea* Escholtz, 1833
Acmaea sp. aff. *A. cerillosensis* Johnson

Refer to *Acmaea cerillosensis* Johnson, 1903, p. 127, Pl. 1, figs. 9, 9a, b.

Material. Two poorly preserved internal molds of small adult shells.

Remarks. *Acmaea cerillosensis* Johnson is characterized by its small size, low conical shell, large apical angle, only slightly coiled, eccentric apex leaning slightly anterior, its circular apertural outline, and very fine growth lines. The poorly preserved specimens from Oklahoma have many of these characters, are about twice the size of Johnson's specimen (possibly a juvenile), and are at least closely related to this form. They are too incomplete for specific determination however.

Stratigraphic and Geographic Position. Middle part of Bridge Creek Limestone Member, Greenhorn Formation, zone of *Mytiloides mytiloides* in hard, ledge-forming, light gray, relatively pure sublithographic limestone (unit 88, sec. 2).

Described Specimens. USNM 172877.

Order MESOGASTROPODA

Mesogastropoda, sp. indet.

Gastropods are generally rare in carbonate rocks of the Colorado Group, partially because of restrictive ecological situations and partially because of the poor condition of preservation in carbonates of organisms with aragonitic shells. Because our knowledge of gastropods in the Graneros, Greenhorn, and Carlile formations is so incomplete, it is worth reporting all occurrences of even poorly preserved specimens.

Two incomplete and generically indeterminate gastropods are known from the lower limestone of unit 81, section 2, upper portion of the lower part of Bridge Creek Limestone Member, Greenhorn Formation, zone of *Sciponoceras gracile*. This zone elsewhere yields the most diverse gastropod assemblage of any Cretaceous chalk or pure limestone unit in the central Western Interior.

One specimen is a small, very low spired, evolute, discoid gastropod generally similar to the Cretaceous genus *Pseudomalaxis* Fischer, 1850 (Family Architectonicidae), or, to a lesser extent, the archeogastropod *Damesia* Holzapfel, 1888 (see Sohl, 1960, Pl. 6, figs. 25–42), but lacking any ornamental characters that might allow specific or generic determination. USNM 242989.

A second specimen is a portion of a smooth ultimate and penultimate whorl of a high-spired gastropod, generally having the shape of an aporrhaid like *Lispodesthes* White, 1876 (1877) (see Stephenson, 1952, Pl. 39, figs. 1–16). USNM 172878.

Figure 5. Diagrammatic representation of open coiled, slightly helical form of *Stomohamites simplex cimarronensis*, n. subsp. (×1); drawn from numerous fragments and imprints. Arrow explained in text.

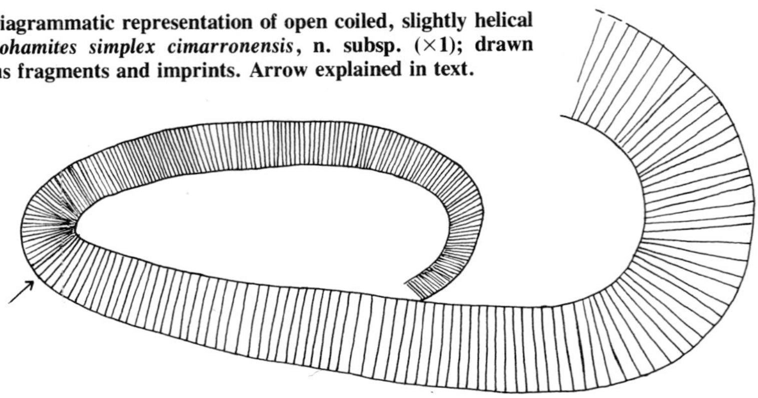

Order CEPHALASPIDEA
Superfamily PYRAMIDELLACEA
Family PYRAMIDELLIDAE

Genus *Pseudomelania* Pictet and Campiche, 1862
Cf. *Pseudomelania* sp.

Material. A single incomplete internal mold.

Description. Small, moderately high spired, with nearly straight sides to the spire and very narrow incised sutures. Whorl flanks very slightly convex. No visible ornament; growth lines very weak, details unknown. Apertural characters and early whorls not preserved.

Remarks. Stephenson (1952, Pl. 35, figs. 42–46) illustrated similar shells and referred them questionably to this genus. His specimens, from the Dexter and Red Branch Members of the Woodbine Formation, are similar to some that have been found at an approximately equivalent level in the Graneros Shale of south-central Colorado in fully marine facies. If this poorly preserved specimen from the Hartland Member is truly a *Pseudomelania*, it extends the known stratigraphic occurrence of the genus in Western Interior marine Cretaceous strata and further documents a partially normal marine range for the genus during at least part of the Cretaceous Period. *Pseudomelania* is generally thought of as an estuarine genus.

Geographic and Stratigraphic Position. Regional zone of *Dunveganoceras albertense*, middle part of Hartland Member, Greenhorn Formation, in grayish-orange, moderately resistant shaly chalk (unit 19, sec. 3).

Figured Specimen. USNM 172879.

Order BASSOMATOPHORA
Superfamily SIPHONARIACEA
Family SIPHONARIIDAE

Genus *Anisomyon* Meek and Hayden, 1860
Anisomyon? sp.

Material. A poorly preserved internal mold of a medium-sized adult shell.

Remarks. Cretaceous species of *Anisomyon* from the Western Interior are characterized by their patelliform shells with ovate apertural section, moderate to high convexity, and asymmetrical apical portion of the shell, which is inclined posteriorly with the apex situated subcentral or in the posterior one-third of the shell length. Fine growth lines and in some cases a few faint radiating ridges characterize the ornament. The specimen from Oklahoma resembles *Anisomyon* in size, shell convexity, the asymmetry of the apical area, and shape of the aperture. The mold also retains faint growth lines and questionably a radiating anterior ridge (possibly a compressional feature). No other characters are preserved; the shell strongly suggests relation to *Anisomyon* but is specifically indeterminate.

Stratigraphic and Geographic Position. In the zone of *Mytiloides mytiloides*, lower portion of the upper part of Bridge Creek Limestone Member, Greenhorn Formation, in hard, light gray, very fine grained, relatively pure limestone (unit 96, sec. 2).

Described Specimen. USNM 172880.

Class CEPHALOPODA
Order AMMONOIDEA de Haan, 1825
Family HAMITIDAE Hyatt, 1900

Genus *Stomohamites* Breistroffer, 1940
***Stomohamites simplex* (d'Orbigny)**
***Stomohamites simplex cimarronensis* Kauffman and Powell, n. subsp.**
(Pl. 9, figs. 1, 3, 4; Figs. 5, 6)

Material. Collections from units 3, 4, and 21, section 3, contain 29 poorly preserved fragments of adult and younger heteromorphs, flattened on slabs of chalky shale and chalky limestone. Shell material is absent, and preservation is by composite molds. One mold shows two partially preserved sutures on a fragment of the phragmocone.

Diagnosis. *Stomohamites simplex cimarronensis*, n. subsp., is a slightly larger and more densely ribbed variety of *S. simplex* (d'Orbigny). Besides the juvenile stage, the adult form contains three complete shafts and at least an open turn of a fourth shaft. Rib size and spacing increases slightly at the beginning of the body chamber.

Description. General form of the adult shell is slightly helical and regularly uncoiled (Fig. 5); coiling in early stages is poorly known, but probably is an open spiral. Adults have three, and part of a fourth, moderately straight nonparallel shafts (Fig. 5). Ornamentation consists of non-tuberculate ribs that are moderately strong, closely spaced, and continuous over the venter where they tend to strengthen slightly (Pl. 9, fig. 1). Ribs are straight and radial except on the tighter turns where they show a distinct aboral slant. The beginning of the body chamber is marked by a slight increase in size and spacing of the ribs (Fig. 5, at arrow; Pl. 9, fig. 4). The aperture is slightly constructed.

Measurements. No attempt has been made to calculate the actual diameter or whorl shape of *S. simplex cimarronensis*, n. subsp., and the measurements shown in Table 11 are all taken from the specimens as they are flattened on bedding planes. As such, the measurements are only approximations of the actual whorl diameter and serve mainly to indicate the size of the subspecies and relate rib counts to ontogeny.

Remarks. The generic assignment for *S. simplex cimarronensis*, n. subsp., is somewhat tenuous because no complete suture lines (see Fig. 6) are preserved to actually identify the large third lateral saddle (Wright, 1957, p. L217) of *Stomohamites* Breistroffer. In addition the characteristic collared aperture of *Stomohamites* was not noticeable except on one crumbled specimen described

Table 11. Summary of Measurements for *Stomohamites simplex cimarronensis* n. subsp.

Specimen	Whorl diameter (mm)	No. of ribs per unit length that is equal to whorl diameter at point of measurement
USNM 167176	4.0	4
USNM 167163	7.4	6
USNM 167162	11.0	8
USNM 167175	26.0	8
USNM 167164	27.0	8
USNM 167160	28.0	8

in the field notes from the type locality. However, material compares so closely to the descriptions and illustrations of *Stomohamites simplex* (d'Orbigny) by Wright (1963, p. 598; Pl. 81, figs. 1a–1c) and Kennedy (1971, p. 67; Pl. 1, figs. 1–8) that we have referred our specimens to that species. Furthermore, the preserved differences between *S. simplex cimarronensis*, n. subsp., and the typical form are more in degree than in kind.

Most notable among these differences is the presence of a third and part of a fourth shaft in the Cimarron County specimens. Fragmentary composite mold preservation on shaly chalk bedding surfaces has not yielded a complete specimen. But the same subspecies is fairly common in the Lincoln-Hartland equivalents in Colorado and Texas. From all of these areas we have been able to observe that the body chamber probably includes all of the least complete shaft and turn, with the last septum occurring near the last part of the penultimate turn (Fig. 5, at arrow).

Stomohamites simplex cimarronensis, n. subsp., differs from the typical form of the species in having three shafts and part of a fourth and in being slightly larger. *S. simplex cimarronensis*, n. subsp., is similar to *S. subvirgulatus* (Spath) in being rather densely ribbed.

Sutures are preserved on one specimen (Fig. 6), but are not well-enough preserved to indicate any one of the hamitid genera. The two sutures illustrated are closely spaced and occur at the position where strengthening of the ribs begins.

The largest available specimens of *S. simplex cimarronensis*, n. subsp., are larger than most of the other described species of *Stomohamites*.

Stratigraphic and Geographic Position. *S. simplex cimarronensis*, n. subsp., occurs in the middle part of the Hartland Member of the Greenhorn Formation in Cimarron County, Oklahoma (units 3, 4, and 21, sec. 3), in the same level at the eastern edge of Eldorado Springs in Boulder County, Colorado, at several other localities in the Western Interior, and at several localities in the lower part of the Eagle Ford Formation in northern Texas. At each of these documented localities, *S. simplex cimarronensis*, n. subsp., occurs with acanthoceratid ammonites and species of *Entolium* and *Inoceramus* described herein.

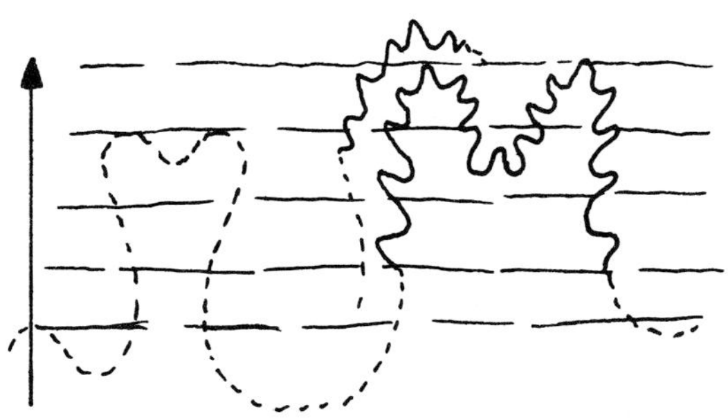

Figure 6. Diagrammatic representation of partially preserved suture pattern of *Stomohamites simplex cimarronensis*, n. subsp. (×4), USNM 167162. Solid lines indicate actual preservation; dashed lines are idealized reconstructions.

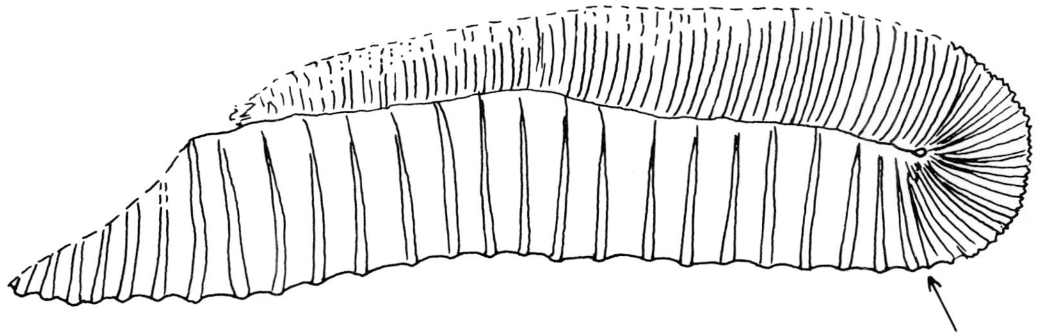

Figure 7. Representation of a partially reconstructed specimen of *Hemiptychoceras* sp. (×4), drawn from USNM 167161 (see also Pl. 8, fig. 5). Arrow explained in text.

Types. Holotype, a flattened fragment of a body chamber, USNM 167160 (collected by Hattin from unit 4, sec. 3 [USGS Mesozoic loc. 30235]; Pl. 9, fig. 1). Paratypes: USNM 167162 (unit 4, sec. 3; Pl. 9, fig. 4); USNM 167164 (unit 4, sec. 3; Pl. 9, fig. 3).

Genus *Hemiptychoceras* Spath, 1925
Hemiptychoceras sp.
(Pl. 9, fig. 5; Fig. 7)

Material. One specimen, a composite mold, is from unit 4, section 3.

Description. The single small specimen shows only the ultimate and penultimate parallel shafts of a small, tightly pressed heteromorph. The shafts are ornamented with radial, nontuberculate ribs that tend to strengthen slightly across the venter. At the beginning of the adult shaft and just past the last turn, the ribs become abruptly coarser (Fig. 7).

Measurements. Length of the larger shaft is 30 mm (Pl. 9, fig. 5). On a specimen showing both the larger and smaller shaft (USMN 167161) the whorl diameters (as preserved) are 5.5 and 3.4 mm, respectively, and each shaft has 5 ribs in a linear distance that is equivalent to its measured whorl diameter.

Remarks. This specimen is assigned to *Hemiptychoceras* Spath because of the abrupt change in rib strength on the last turn (Wright, 1957, p. L217). W. A. Cobban of the U.S. Geological Survey kindly provided a photograph of a specimen that has been named *Hemiptychoceras reesidei* Cobban and Scott (1972, p. 45). The *H. reesidei* specimen is from the Bridge Creek Limestone Member of the Greenhorn Formation in Colorado. Our specimen may be conspecific, but somewhat older.

Stratigraphic and Geographic Position. *Hemiptychoceras* sp. occurs in the middle part of the Hartland Member of the Greenhorn Formation, unit 4, section 3.

Type. Hypotype, a composite mold, USNM 167161 (unit 4, sec 3; Fig. 7; Pl. 9, fig. 5).

Family SCHLOENBACHIIDAE Parona and Bonarelli, 1897

Genus *Quitmaniceras* Powell, 1963
Quitmaniceras sp. cf. *Q. brandi* Powell, 1963a
(Pl. 10, fig. 5)

Quitmaniceras brandi Powell, 1963a, Jour. Paleontology, v. 37, p. 314, Pl. 32, figs. 6, 8, 11, 12, 14–16; Figs. 3i, 3p, 3q.

Material. Three fragmental specimens from unit 86, section 2. The ammonites are preserved in hard chalky limestone as internal molds, without suture lines.

Description. The shell is small (as much as 30 mm in diameter), moderately evolute, slightly compressed, and costate. Ribs are strong and nearly straight on flanks becoming slightly weaker and projected slightly forward across the venter. Both primary and secondary ribs are present. Secondaries disappear near midflank. Both rib types are irregularly turberculate, with weak single ventrolateral and siphonal tubercles. Bullate swellings near umbilicus are noticeable on one specimen but not on USNM 167171 (Pl. 10, fig. 5). The venter, preserved on one specimen, contains what seems to be a weak keel, as illustrated by Powell (1963a, Pl. 32, figs. 6 and 14).

Stratigraphic and Geographic Position. *Quitmaniceras* sp. cf. *Q. brandi* is associated with *Allocrioceras*?, *Worthoceras*, and *Mytiloides mytiloides* Mantell in the upper part of the exposed Bridge Creek Limestone Member, Greenhorn Formation. This distinctly Turonian fauna resembles that of the type occurrence of *Q. brandi* Powell in northern Chihuahua, Mexico, where vascoceratids and mammitids occur along with *M. mytiloides*. Vascoceratids have been reported in association with *M.* "*labiatus*" by Cobban and Reeside (1952, p. 1018), Cobban and Scott (1972, p. 83), and Hattin 1975 from the upper part of the Greenhorn Formation in southeast-central Colorado and western Kansas.

Figured Specimen. Internal mold of a small fragment, probably a body chamber, USNM 167171 (unit 86, sec. 2).

Family BACULITIDAE Meek, 1876

Genus *Sciponoceras* Hyatt, 1894
Sciponoceras gracile (Shumard), 1860
(Pl. 10, fig. 4)

Baculites gracilis Shumard, 1860, p. 596.

B. gracilis Shumard. Stanton, 1893 (1894), p. 166.

B. gracilis Shumard. Adkins, 1928, p. 206.

Sciponoceras gracile (Shumard), Cobban and Reeside, 1952, p. 1017. Cobban and Scott, 1972, p. 47, Fig. 18; Pl. 17, figs. 9-29. Hattin, 1975, Pl. 6, figs. A, B, N.

Material. Two lots contain three specimens from units 78 and 81, section 2, preserved in hard chalky limestone.

Description of USNM 167172. Shaft is slender and straight, with prominent rounded prosiradiate ribs and constrictions. Diameter of the shaft ranges from 7 to 9 mm along its 75-mm length. An estimated 19 ribs occur over the total length of specimen. Toward the more mature end of the shaft, the ribs are less prominent and constrictions are more shallow. The sutures are typical for that species.

Stratigraphic and Geographic Position. *Sciponoceras gracile* (Shumard) is the index species for an extremely widespread ammonite zone (Cobban and Reeside, 1952, p. 1017). *S. gracile* commonly occurs with other equally widespread ammonites such as *Metoicoceras whitei* Hyatt, *M. acceleratum* Hyatt, *Worthoceras* spp., *Kanabiceras septemseriatum* (Cragin), *Allocrioceras* spp., *Proplacenticeras* spp., *Scaphites* spp., and *Borissjakoceras* sp. The geographic range of general elements of this ammonite fauna is from Alaska to Mexico with *Sciponoceras* occurring commonly from Montana to Texas.

In Cimarron County, Oklahoma, *S. gracile* occurs in units 78 and 81, section 2.

Figured Specimen. Internal mold of an entirely septate fragment, 75 mm long; USNM 167172 (unit 81, sec. 2).

Family DESMOCERATIDAE Zittel, 1895

Genus *Desmoceras* Zittel, 1884
Desmoceras?, s. l., sp.
(Pl. 9, figs. 2, 6)

Remarks. Numerous poorly preserved imprints and crushed fragments of composite molds are questionably referred to *Desmoceras* Zittel, s. l. All of these are from chalky limestone units 4 and 13, section 3. One small specimen (USNM 167166; Pl. 9, fig. 7) is probably a juvenile of *Pseudouhligella* that is found commonly in Gulf Coast rocks of Cenomanian age. The remaining specimens (including USNM 167169 and USNM 167165; Pl. 9, figs. 2 and 6, respectively) have the following characteristics that allow their placement in the genus *Desmoceras*, s. l.: The shell is moderately inflated to slightly compressed and very involute; it reaches a maximum diameter of about 50 mm. Weak, nearly straight to slightly falcoid, low ribs are without tubercles. On some specimens, constrictions and an associated bordering rib occur at a frequency of 5 to 8 per volution. None of the specimens shows any indication of a keel, or any definite example of ribs crossing the venter. The venter is not well preserved but seems to be arched. A specimen comparing closely with *Desmoceras* s. l. of Cimarron County has been recorded in the Lincoln Member of central Kansas.

Figured Specimens. Composite molds of two specimens; USNM 167169 (unit 4, sec. 3); USNM 167165 (unit 13, sec. 3).

Subgenus *Pseudouhligella* Matsumoto, 1942
***Pseudouhligella* sp.**
(Pl. 9, fig. 7)

Remarks. One small specimen (estimated diameter, 15 mm) is represented by a partial imprint in chalky limestone, unit 4, section 3. A rubber cast (Pl. 9, fig. 7) of the imprint reveals a grounded whorl with constrictions and no other ornament. The constrictions are projected forward across the broadly rounded venter. Degree of involution is not observable, nor is any suture preserved. The specimen is associated with numerous other similar forms that are more ornamented, having weak, nearly straight ribs and constrictions. The latter group is referred questionably to *Desmoceras*, s. l.

Figured Specimen. Rubber cast of imprint (unit 4, sec. 3).

Family ACANTHOCERATIDAE Hyatt, 1900
Subfamily MANTELLICERATINAE Hyatt, 1903

Genus *Pseudocalycoceras* Thomel, 1969
***Pseudocalycoceras* spp. 1 and 2**
(Pl. 10, fig. 1)

Remarks. Eighteen poorly preserved ammonites are referred to *Pseudocalycoceras* spp. 1 and 2. Occurrence in Cimarron County is in the upper parts of the Lincoln and Hartland Members. All of the specimens are preserved without shell material and include crude composite molds and fragments of internal molds. Members of this group are referred to *Pseudocalycoceras* on the basis of costation and apparent whorl shape. The ammonites are high-whorled and coarsely ribbed and maintain tubercules on the ribs to a late stage of ontogeny. Otherwise, the costation is reminiscent of both *Calycoceras* Hyatt and *Tarrantoceras* Stephenson.

From units 20, 22, and 24, section 2, 11 specimens (including USNM 167173; Pl. 10, fig. 1) were collected that resemble some specimens of *Tarrantoceras rotatile* Stephenson and *T. lillianense* Stephenson (1955, Pl. 5), as well as some unfigured topotypes of these species. The similarities include greatly reduced inner ventrolateral tubercles and moderately weak, conical outer ventrolateral and siphonal tubercles. However, the coarseness of the costation and the distinct aboral slant of the ribs suggest a relationship to the genus *Pseudocalycoceras* Thomel (1969, p. 650). The venter is partially preserved on one specimen and shows three rows of closely spaced tubercles having weak connecting ribs crossing the venter. Umbilical ends of the ribs indicate nothing conclusive about the presence or shape of possible umbilical tubercles. Bits of suture preserved on one specimen give no hint of the distinctive tarrantoceroid pattern described by Stephenson (1955, p. 59). The present group of specimens is therefore referred to *Pseudocalycoceras*.

Two "species" are recognized based on stratigraphic separation, size and variation in length and strength of ribbing. *Pseudocalycoceras* sp. 2 is not illustrated. It is distinguished from *P.* sp. 1 in having stronger, straighter ribs on fragments of the adult(?) living chamber.

Stratigraphic and Geographic Position. *Pseudocalycoceras* sp. 1 occurs in the upper part of the Lincoln Member, Greenhorn Formation, in units 20, 22, and 24, section 2. *Pseudocalycoceras* sp. 2 occurs in the upper part of the Hartland Member, Greenhorn Formation, in unit 58, section 2, and in unit 13, section 3.

Figured Specimen. Fragment of an internal mold, *Pseudocalycoceras* sp. 1, USNM 167173 (unit 22, sec. 2).

Subfamily ACANTHOCERATINAE Hyatt, 1900

Genus *Acanthoceras* Neumayr, 1875

Numerous small internal and composite molds in the soft shaly chalk and chalky limestone of section 3 are referred to *Acanthoceras* and are distinguished mainly on characteristics of ribbing and ventral ornamentation. Two morphologic groups of this genus are recognized, but neither group is referred to any species. All of the specimens appear to be juveniles and measure less than 30 mm in diameter. More than one genus may be represented within these groups. Nevertheless, these ammonite groups are widely recognized in the Hartland Member of the Greenhorn Formation, or the equivalent interval, in Colorado, Kansas, and adjacent areas. Consequently, these forms deserve description as a significant part of the total fauna.

Generally, all of these ammonites are costate, having straight to slightly flexuous ribs. Primary and secondary ribs occur. Primary ribs lack umbilical tubercles and do not reach the umbilical shoulder. The whorls appear to be square or slightly compressed, and ventral impressions indicate a slight forward projection of the ribs.

Both morphologic groups of acanthoceratids described here occur at several places in the Western Interior. The middle Hartland ammonite bed in Boulder County, Colorado, contains representatives of both groups. Group "B" also occurs in the lower part of the Britton Member of the Eagle Ford Formation in north Texas. The Hartland Member ammonite fauna, although it has not yet been found well preserved, seems to be a widespread and easily recognizable biostratigraphic marker.

Acanthoceras Group "A"
(Pl. 10, fig. 3)

Material. Four specimens, from units 1 and 2, section 3, are weak composite molds in soft shaly chalk, affording only lateral views of the ornamentation.

Description. The specimens representing *Acanthoceras*? group "A" are densely costate with fine, sharply defined moderately biconcave ribs that are projected forward at the ventrolateral margin. The rib pattern is somewhat irregular in that from none to three or four secondary ribs occur between primary ribs. Secondary ribs disappear near midflank, and contain only weak ventrolateral tubercles. Umbilical and inner ventrolateral tubercles are radially elongate. Poorly preserved outer ventrolateral tubercles appear to be more spinose. Siphonal tubercles are not preserved. At a diameter of 23 mm, one specimen (USNM 167174) contains about 28 primary ribs and 22 secondary ribs per whorl.

Remarks. Other generic names could be applied to this group, especially with regard to the extremely fine ribbing, but with such fragmental materal, little would be gained by further comparisons.

Figured Specimen. Composite mold of a small fragment preserved in soft shaly chalk, USNM 167174, from unit 2, section 3.

Acanthoceras Group "B"
(Pl. 10, figs. 2, 6, and 7)

Material. Twelve specimens of small ammonites from units 2, 3, and 4, section 3. Occurrence is both in soft shale (Pl. 10, fig. 7) and chalky limestone (Pl. 10, figs. 2 and 6).

Description. Group "B" is coarsely costate with numerous rounded ribs of variable strength. Ribs are mostly straight to weakly sinuous and slant slightly forward. Poorly preserved ventral composite molds indicate that the ribs are discontinuous across the venter, but the position of the siphonal tubercle suggests little or no forward projection. Both primary and secondary ribs occur. Primary ribs on most specimens are effaced at midflank and reappear at the umbilicus as radially elongate to slightly curved, strong umbilical tubercles. One specimen (USNM 167168) is an internal mold and shows only faint ribs and spiny tubercles (Pl. 10, fig. 2). Ventrolateral tubercles are conical to spinose, whereas the siphonal tubercles appear to be slightly clavate to conical.

Remarks. *Acanthoceras* group "B" is comparable to the most juvenile whorls of *Acanthoceras barcusi* Jones (Late Cenomanian) of northern Mexico and to fine-ribbed forms of *Protacanthoceras* Spath. The specimen (Pl. 10, fig. 2) is the largest and best preserved of several other small, unnumbered specimens. Forms closely resembling this specimen are known from the Greenhorn Formation elsewhere in the Western Interior, but so far no mature specimens have been found that indicate a specific affinity. The other specimens (Pl. 10, figs. 6, 7) are somewhat larger and very much like a plaster cast of a specimen collected by W. A. Cabban from the Belle Fourche Shale of Wyoming. The three rows of closely spaced ventral clavi of *Protacanthoceras* are not distinctly indicated on any of these specimens, and we feel that until better material becomes available, *Acanthoceras* is the best generic category of this group.

Measurements. The specimens range in maximum diameter from 12 to 26 mm. Most of the specimens are from 19 to 23 mm in diameter. Estimated rib counts on three figured specimens are as follows: USNM 167167, 23 primary ribs, and 8 secondary ribs at 26-mm diameter; USNM 167168, 16 primary ribs and 17 secondary ribs at 19-mm diameter; and USNM 167170, 18 primary ribs and 9 secondary ribs at 22-mm diameter.

Figured Specimens. All are composite molds: USNM 167170 (unit 2, sec. 3), USNM 167168 (unit 3, sec. 3) USNM 167167 (unit 4, sec. 3), USNM 167168 (unit 4, sec. 3).

Phylum ARTHROPODA
Subphylum CRUSTACEA
Class CIRREPEDIA

Acrothoracican Barnacles

Wedge- and teardrop-shaped borings in *Inoceramus* shells comparable in size, shape, and distribution to those of acrothoracican barnacles (Seilacher, 1969, Figs. 5, 6) occur widely in the Cretaceous strata of the Western Interior. In Oklahoma these were noted in shells from the middle part of the Hartland (unit 52, sec. 2) and middle part of the Bridge Creek Members (unit 86, sec. 2), USNM 172881 and 172882, respectively.

Notes on Trace Fossils

No attempt was made to study, in depth, the very numerous and diverse trace fossil assemblage in the Dakota Group, Graneros Shale, and Greenhorn Formation of Oklahoma. Frey (1970) and Hattin (1971) illustrated the kind of diversity that might be expected in the carbonate units from their studies of the Fort Hays Limestone Member of the Niobrara Formation and Greenhorn Formation, respectively, in Kansas. An even greater diversity of types occurs in Dakota Group and Graneros Shale sandstone and siltstone facies. Five types of trace fossils were noted consistently in the thicker bedded, fine-grained units of the Greenhorn Formation during this research, and specimens were collected for a subsequent study of the trace fossils throughout the Colorado Group. They are mentioned here as principal parts of the Greenhorn macrofossil assemblages.

The principal form is an irregularly curved, single or bifurcating ribbon burrow, presumably that of a worm, identical to those called *Planolites* sp. by Frey from the Fort Hays Limestone Member (Frey, 1970, Pl. 5, figs. 1, 3) and by Hattin from the Greenhorn Formation (Hattin, 1971, Fig. 11). These occur in the limestone beds of the Bridge Creek Member and in harder blocky chalk beds of the Hartland Member.

Three types of small vertical or nearly vertical, simple to simply branching tubes were identified in limestones of the Bridge Creek: (1) *Chondrites* sp., such as those illustrated by Frey from the Fort Hays (1970, Pl. 3, fig. 4A; Pl. 5, fig. 3A) and by Hattin from the Greenhorn (1971, figs. 12, 13); (2) small, very narrow, slightly sinuous tubes commonly filled with limonite and in some cases associated with pyrite, closely resembling *Trichichnus* sp. of Frey (1970, Pl. 6, figs. 5, 7) in size and shape; and (3) rare vertical tubes, 1 to 3 mm in diameter and lined with an inner sediment tube, which were illustrated by Frey (1970, Pl. 8, fig. 5) and can be referred generally to the trace fossil genus *Tigillites*.

Finally, horizontal to slightly inclined U-shaped burrows are moderately abundant in harder chalk units and limestones of the Hartland and Bridge Creek Members. Although no spreite were observed within the "U" and preservation was too poor to determine the presence or absence of scratch marks on the burrow walls, in size, shape, and orientation these structures resemble closely arthropod burrows referred to *Rhizocorallium* sp. (see Häntzschel, 1962, Pl. W207, fig. 2) and are here included in this form-genus.

Phylum PROTOZOA
Class RHIZOPODA
Order FORAMINIFERIDA Eichwald, 1830
Suborder ROTALIINA Delage and Herouard, 1896
Superfamily GLOBIGERINACEA Carpenter; Parker and Jones, 1862
Family HETEROHELICIDAE Cushman, 1927

Genus *Heterohelix* Ehrenberg, 1843
***Heterohelix globulosa* (Ehrenberg)**
(Pl. 11, fig. 5)

Textularia globulosa Ehrenberg, 1840, p. 135, Pl. 4, figs. 1B, 2B, 4B, 5B, 7B, 8B. For recent synonymy, see Eicher and Worstell, 1970, p. 29b.

Lectotype. Selected from syntypic series by Pessagno (1967, p. 260); the specimen figured by Ehrenberg (1840, Pl. 4, fig. 5B).

Material. Numerous small, moderately well preserved individuals occur throughout the sampled interval above unit 10, section 2. Tests are mostly calcite filled, with shell surfaces slightly etched by corrosion that has obliterated any original ornamentation. The specimens range in length from 0.10 to 0.39 mm. Length of the figured specimen is 0.22 mm.

Diagnosis. Lateral outline of test variable, ranging from broadly to narrowly flaring with greatest width near apertural end; aperture low and smooth, commonly with a discontinuous lip; chambers are marked by very fine longitudinal ornament.

Remarks. Cushman (1946, p. 105, 106, Pl. 45, figs. 9-15) and others (for example, Pessagno, 1967) have used the name *Heterohelix globulosa* for very late Cretaceous (Campanian and Maestrichtian) forms in the Gulf of Mexico area. In Cenomanian-Turonian faunas, this species name is also commonly applied in the Western Interior of the United States in reference to nearly all typically biserial heterohelicids with globular chambers, with or without costae (for example, Eicher and Worstell, 1970, p. 296, Pl. 8, figs. 3-8). Surficial ornament is not well-enough preserved on any specimens to permit detailed comparisons with other species. The shape range of the tests and the variable amount of chamber overlap in our specimens include those that could be referred to *H. moremani* (Cushman) and *H. reussi* (Cushman). Our specimen (Pl. 11, fig. 5) is more nearly like *H. moremani* than *H. globulosa* as figured by Pessagno (1967, Pl. 87) from Campanian strata of Texas. Until this group of closely related heterohelicids is treated in detail through a population systematics approach, the accepted lumping of several species (possibly variants) under *H. globulosa* in the Western Interior will necessarily be followed.

Stratigraphic and Geographic Position. This species is widespread in Upper Cretaceous rocks of the Western Interior of the United States, the foothills area of Alberta, Canada (Wall, 1967), and in Alaska (Tappan, 1962). In Cimarron County, Oklahoma, the species first occurs in unit 10, section 2, and is common to abundant in all higher samples of the Lincoln, Hartland, and Bridge Creek Limestone Members of the Greenhorn Formation.

Figured Specimen. A moderately well preserved individual, USNM 689375 (unit 44, sec. 2).

Family SCHACKOINIDAE Pokorny, 1958

Genus *Schackoina* Thalmann, 1932

Both *Schackoina cenomana* (Schacko) and *S. multispinata* (Cushman and Wickenden) occur discontinuously within the Lincoln and Hartland Members in Cimarron County. Poor preservation of these delicate forms precludes their illustration here, but good descriptions and illustrations of typical Greenhorn forms of both species are found in Eicher and Worstell (1970, Pl. 9), and in Loeblich and Tappan (1961, Pl. 1).

Although noted less commonly than the more robust foraminifers, *Shackoina cenomana* and *S. multispinata* occur in units 38 and 44, section 2. Their absence elsewhere in beds of the Lincoln and Hartland Members of the Greenhorn is attributed to poor preservation factors. These two species of *Schackoina* occur commonly in the lower (Late Cenomanian) part of the Eagle Ford Formation in Texas.

Family PLANOMALINIDAE Bolli, Loeblich, and Tappan, 1957

Genus *Globigerinelloides* Cushman and Ten Dam, 1948
Globigerinelloides bentonensis (Morrow)
(Pl. 11, figs. 1, 4)

Anomalina bentonensis Morrow, 1934, p. 201, Pl. 30, fig. 4. For recent synonymy, see Eicher and Worstell, 1970, p. 297.

Material. Numerous well-preserved specimens are available from all sampled units in the Lincoln and Hartland Members of the Greenhorn Formation. Tests are mostly corroded, but are best preserved in units 34 and 44, section 2. The diameter range of all specimens is from 0.18 to 0.50 mm. The illustrated specimens (Pl. 11, figs. 1, 4) are 0.38 and 0.35 mm in diameter, respectively.

Diagnosis. Not quite symmetrically planispiral; moderately involute; chambers regularly inflated and rapidly expanding; periphery flattened to broadly arched on latest chambers, with last one or two chambers commonly wider than high.

Remarks. This species has been described adequately by Loeblich and Tappan (1961, p. 267) and by Eicher (1965, p. 904). Where only small specimens are available, *Globigerinelloides bentonensis* is sometimes confused with its smaller and more evolute relative, *G. caseyi* (Bolli, Loeblich, and Tappan).

Forms occur (Pl. 11, fig. 1) that have inflated chambers as in *G. bentonensis* but are also more evolute, as in *G. caseyi*. Nevertheless, no real intergradation seems to occur between these two species in Cimarron County.

Stratigraphic and Geographic Position. This species, along with *Globigerinelloides caseyi* (not figured), is common in the upper Cenomanian rocks of the southern part of the Western Interior and in Texas. The lowest occurrence of *G. bentonensis* in Cimarron County is in unit 19, section 2, where the form is rare. It first occurs in abundance in the sample from unit 23, section 2, where *G. caseyi* first appears along with flood abundance of *Clavihedbergella simplex* (Morrow). *Globigerinelloides bentonensis* occurs commonly through unit 75, section 2 and throughout the sequence at section 3. Its stratigraphic occurrence is about equal to that of *Rotalipora greenhornensis* (Morrow). This generality also holds in the Eagle Ford Formation in Texas.

Figured Specimens. Well-preserved individuals: USNM 689376 (unit 44, sec. 2; Pl. 11, fig. 1), USNM 689377 (unit 44, sec. 2; Pl. 11, fig. 4).

Globigerinelloides caseyi (Bolli, Loeblich, and Tappan)

Planomalina caseyi Bolli and others, 1957, p. 24, Pl. 1, figs. 4, 5. For synonymy, see Pessagno, 1967, p. 276.

Material. Uncommon, small individuals occur in most samples above unit 23, section 2, throughout the higher parts of the Lincoln Member and the entire Hartland Member. Most of the specimens were broken during sample reduction or were too badly corroded to permit illustration.

Diagnosis. Small, slightly asymmetrical, slightly involute to evolute; chambers not greatly inflated or rapidly expanding in the last whorl; periphery somewhat more labate than in *Globigerinelloides bentonensis*.

Remarks. *G. caseyi* is well illustrated and described by Loeblich and Tappan (1961, p. 268, Pl. 2, figs. 3a–7b) as *G. eaglefordensis* (Moreman). The present species is not easily separated from *G. bentonensis* (Morrow) when only juveniles or incomplete specimens are available.

Stratigraphic and Geographic Position. *G. caseyi* is recorded first in unit 23 (lower Lincoln Member), section 2. It co-occurs with *Globigerinelloides bentonensis* in all later sampled units of the Lincoln and Hartland Members of the Greenhorn Formation. *G. caseyi* is nearly always present in plankton-rich beds in the Greenhorn of Kansas and Colorado and in the Eagle Ford Formation of Texas. Forms referable to this species also occur in the Grayson Formation (Lower Cenomanian) in Texas (Loeblich and Tappan, 1961) and in beds of Coniacian age in Texas.

Family ROTALIPORIDAE Sigal, 1958
Subfamily HEDBERGELLINAE Loeblich and Tappan, 1961

Genus *Hedbergella* Bronnimann and Brown, 1958
Hedbergella amabilis Loeblich and Tappan
(Pl. 11, fig. 6)

Hedbergella amabilis Loeblich and Tappan, 1961, p. 274, Pl. 3, figs. 1–10. For synonymy, see Pessagno, 1967, p. 281, Pl. 3.

Material. This species is present in considerable numbers in all samples where foraminifera were recovered in the Lincoln, Hartland, and Bridge Creek Limestone Members of the Greenhorn Formation. Most of the specimens are small with corroded surfaces. The tests are mostly filled with calcite. The greatest diameter of the figured specimen is 0.33 mm; other specimens range up to 0.37 mm in diameter.

Remarks. Eicher and Worstell (1970, p. 300) pointed out that *Hedbergella amabilis* grades morphologically into both *Clavihedbergella simplex* (Morrow) and *H. delrioensis* (Carsey). The same gradation seems to occur in Cimarron County with *H. amabilis* being somewhat less abundant than the other two members of the group. However, the relative numbers of these three species are not constant from sample to sample through the vertical sequence and, therefore, probably were not constant through time. Careful analysis of new data on these species from several areas might suggest a discernible selection pressure resulting from environmental circumstances.

Until population studies are available, *Hedbergella amabilis* is best considered a form intermediate between *H. delrioensis* and *C. simplex*, although we do not believe that these three species form a phyletic lineage. For other views, see Douglas (1969).

Stratigraphic and Geographic Position. *Hedbergella amabilis* is widespread in the southern part of the Western Interior, Texas, and Mexico, in Upper Cenomanian and Lower Turonian rocks. This form occurs in abundance in units 19 through 37 and is common above unit 38 in section 2 and throughout section 3. No occurrence of *H. amabilis* was noted below the Lincoln Member in Cimarron County.

Figured Specimen. A well-preserved individual, USNM 689378 (unit 34, sec. 2).

Hedbergella delrioensis (Carsey)
(Pl. 11, fig. 3)

Globigerina cretacea d'Orbigny var. *delrioensis* Carsey, 1926, p. 43. For synonymy, see Eicher and Worstell, 1970, p. 302.

Material. This species is present in the upper part of the upper shale member of the Graneros Shale and is abundant in all sampled beds throughout the Greenhorn Formation. The best preservation is in unit 34 where large specimens are numerous. Maximum diameter noted for the species in all samples is 0.38 mm, with the figured specimen (Pl. 11, fig. 3) having a diameter of 0.37 mm.

Remarks. *Hedbergella delrioensis* is an important species, as it seems to be the basic stock from which all of the other Cenomanian and Turonian hedbergellan species evolved. Pessagno (1967, p. 282-283, 298) and Eicher and Worstell (1970, p. 302) presented phylogenetic views that are accepted here. Forms transitional between *H. delrioensis* and *H. portsdownensis* are numerous throughout a size range from 0.40 mm to 0.58 mm in the stratigraphic range of the two species in Cimarron County.

Preservation is not good enough in most instances to study the details of the apertural flap in *Hedbergella delrioensis*. Where this structure is found well preserved, there is no hint of a serrated distal margin (similar to that of *H. amabilis* as illustrated by Loeblich and Tappan, 1961, Pl. 3), which has been observed in *H. delrioensis* from the Eagle Ford Formation in north Texas.

Stratigraphic and Geographic Position. This species is commonly present in the Graneros Shale (units 10 to 17, sec. 2), but is unusually small. In the Greenhorn Formation, *H. delrioensis* is abundant in all samples from each section. The species is widespread throughout the Western Interior from Canada to Mexico and throughout the Gulf Coast region, ranging from Early Cenomanian through Turonian and perhaps as late as Early Campanian (Pessagno, 1967, p. 283).

Figured Specimen. A well-preserved individual, USNM 689379 (unit 34, sec. 2).

Hedbergella planispira (Tappan)

Globigerina planispira Tappan, 1940, p. 122, Pl. 19, fig. 12. For synonymy, see Eicher and Worstell, 1970, p. 302.

Material. This small species is present in moderate numbers in all samples of the Greenhorn Formation and is present in the upper shale member of the Graneros

Shale. The specimens are mostly corroded and generally too poor for illustration.

Diagnosis. Test small, a slightly trochoid spiral of 2 to 2.5 whorls, with 5 to 7 chambers in the last whorl; spiral side is not convex.

Remarks. This species is not easily separated from *Globigerinelloides* owing to similarities in size and chamber pattern on the umbilical side of the test. As Eicher and Worstell (1970, p. 298, 304) pointed out, the species are best distinguished by observing the distinctly trochoid spiral side of *Hedbergella planispira*. In the Greenhorn fauna, no other species closely resemble *H. planispira*; discussions of relationships with other faunas are in Bandy (1967), Pessagno (1967), Douglas (1969), Hermes (1969), and Eicher and Worstell (1970).

Stratigraphic and Geographic Position. This species occurs throughout the sequence in Cimarron County from the upper Graneros Shale to the middle of the Bridge Creek Limestone Member of the Greenhorn Formation. It is also widely known in the remainder of the Western Interior and in the Gulf Coast area.

Hedbergella portsdownensis (Williams-Mitchell) (Pl. 12, figs. 7, 12)

Globigerina portsdownensis Williams-Mitchell, 1948, p. 96, Pl. 8, fig. 4. For synonymy, see Eicher and Worstell, 1970, p. 304.

Material. This species is common to abundant in the Greenhorn Formation, where most specimens are broken and corroded or are encrusted with chalky calcification as in samples from the Bridge Creek Limestone Member. Some specimens range up to 0.55 mm in diameter. The figured individuals (Pl. 12, figs. 7, 12) have a greatest diameter of 0.47 and 0.45 mm, respectively, and are both from unit 93, section 2.

Diagnosis. Test of moderate to large size, ornate, with strong spiral convexity; consisting of about 3 whorls, each containing 4.5 to 6 globular chambers; the last chamber commonly appears displaced slightly toward the umbilicus; periphery strongly lobate with chambers increasing moderately in size.

Remarks. Several recent workers (Bandy, 1967; Pessagno, 1967; Eicher and Worstell, 1970) have suggested that *Hedbergella brittonensis* Loeblich and Tappan (1961) is a subjective junior synonym of *H. portsdownensis*. This view is followed here. The intergradational relationship between *H. portsdownensis* and *Whiteinella aprica* (Loeblich and Tappan) has already been indicated (Eicher and Worstell, 1970, p. 314). The same gradation is evident in the upper part of the Hartland Member in units 50 to 68 of section 2, where intermediate stages between the two species are present. But only in unit 75 of section 2 (top unit of Hartland Member) does the *W. aprica* form appear. In the Eagle Ford Formation in north Texas the same pattern of evolution occurs within a correlative interval. In spite of the apparent complete gradation between *W. aprica* and *H. portsdownensis*, both forms are considered taxonomically valid.

Stratigraphic and Geographic Position. The first appearance of *Hedbergella portsdownensis* in Cimarron County is in unit 19, section 2, near the base of the Lincoln Member of the Greenhorn Formation. It ranges upward through the Greenhorn, reaching maximum abundance and size in units 93, 97, and 99 of the same section. This species co-occurs with *H. delrioensis* in the youngest part of the range zone of the latter.

Figured Specimens. USNM 689380 (unit 93, sec. 2; Pl. 12, fig. 7), USNM 689381 (unit 93, sec. 2; Pl. 12, fig. 12).

Genus *Clavihedbergella* Banner and Blow, 1959
Clavihedbergella simplex (Morrow)
(Pl. 11, fig. 2)

Hastigerinella simplex Morrow, 1934, p. 198, Pl. 30, fig. 6. For synonymy, see Pessagno, 1967, p. 285.

Material. A few broken, or otherwise poorly preserved specimens are available from samples of the uppermost unit of the Graneros Shale through the top of the Hartland Member of the Greenhorn Formation. The best preservation of this species is in unit 44 of section 2. The figured specimen has a greatest diameter of 0.33 mm, with other specimens reaching a maximum measured diameter of 0.42 mm.

Diagnosis. Test more trochospiral than other species of the genus; contains 2 distinct whorls, usually with 4 chambers in the last whorl; strongly quadrilobate with the last 2 chambers increasingly elongate radially.

Remarks. This species is basically hedbergellan with radially elongate final chambers. As such it is closely similar to *Hedbergella amabilis* Loeblich and Tappan. Pessagno (1967, p. 285) and Eicher and Worstell (1970, p. 306) have adequately discussed this relationship.

Stratigraphic and Geographic Position. *Clavihedbergella simplex* first appears in Cimarron County in unit 17, section 2, upper part of the Graneros Shale. Although Eicher (1966) reported *C. simplex* from as high as the middle part of the Carlile Shale in Colorado, the species was not recognized above the base of the Bridge Creek Limestone Member of the Greenhorn Formation in section 2. This species occurs in the Lincoln–Hartland–Bridge Creek equivalents in the Eagle Ford Formation in north Texas and in western Texas and northern Mexico.

Figured Specimen. A moderately well preserved individual, USNM 689382 (unit 44, sec. 2).

Genus *Praeglobotruncana* Bermudez, 1952
Praeglobotruncana stephani (Gandolfi)

Globotruncana stephani Gandolfi, 1942, p. 130, Pl. 3, figs. 4, 5; Pl. 4, figs. 36, 37, 41–45; Pl. 6, figs. 4, 6; Pl. 9, figs. 5, 8; Pl. 13, fig. 5; Pl. 14, fig. 2.

Material. The few specimens that are well-enough preserved to identify come from unit 75 of section 2.

Remarks. Specimens of this species were not found sufficiently well preserved to permit illustration, but they conform to the descriptions by Loeblich and Tappan (1961), Pessagno (1967), and Eicher and Worstell (1970), who have adequately illustrated and discussed this form. Where preservation is poor, *Praeglobotruncana stephani* is difficult to distinguish from immature specimens of *Marginotruncana difformis* (Gandolfi), which, in its lower stratigraphic range, co-occurs with *P. stephani*.

Stratigraphic and Geographic Position. In Cimarron County, *Praeglobotruncana stephani* was not found below unit 75, section 2, and occurred only questionably in the Bridge Creek Limestone Member at the same locality. Other workers have reported this form as ranging from the Graneros Shale upward into the Carlile Shale (Eicher and Worstell, 1970) in the Western Interior and from equivalent rocks in Texas and Mexico (Pessagno, 1967; Powell, 1970).

Subfamily ROTALIPORINAE Sigal, 1958

Genus *Rotalipora* Brotzen, 1947
Rotalipora cushmani (Morrow)
(Pl. 12, figs. 5, 6)

Globorotalia cushmani Morrow, 1934, p. 199, Pl. 31, figs. 2, 4. For synonymy, see Pessagno, 1967, p. 292.

Material. Large and moderately well preserved specimens are common in unit 75 of section 2 and are sporadically present in most samples of the Lincoln and Hartland Members. The specimens (Pl. 12, figs. 5, 6) are 0.70 and 0.63 mm in diameter, respectively. The largest specimens measured reached a diameter of 0.72 mm.

Diagnosis. A large, biconvex trochospiral form with acute, lobate periphery; 4 to 6 chambers in the last whorl, with each chamber having a thickened, imperforate keel; ventral side with sutural supplementary apertures extending about halfway to the periphery.

Remarks. The considerable morphologic variation throughout the geographic range of *Rotalipora cushmani* is adequately illustrated and described by Loeblich and Tappan (1961, p. 297, 298; Pl. 8, figs. 1-10). This species reaches its maximum development in the upper part of the Hartland Member of the Greenhorn, where numerous unusually large specimens are common. The illustrated forms represent the typical form of the species, although variants occur that have less prominent keels and chambers more inflated on the spiral side. To date, these variations are not known to have stratigraphic value.

Stratigraphic and Geographic Position. The lowest occurrence of *Rotalipora cushmani* in Cimarron County is in unit 34 of section 2 where the species is rare. The uppermost sample containing *R. cushmani* is from unit 75 of the same section. The species is rare in the Lincoln Member and in the lower part of the Hartland Member, becoming abundant above unit 64 of section 2. Eicher and Worstell (1970) reported *R. cushmani* in the lower part of the Bridge Creek Limestone Member at several localities in Colorado and Kansas. Poor recovery and preservation prevented adequate sampling of the foraminifer assemblage in the lower 11 m (35 ft; units 76 through 92) of the exposed Bridge Creek in Cimarron County. This condition has probably caused the noted upward termination of the ranges of several species to be unreliable.

Figured Specimens. USNM 689383 (unit 75, sec. 2; Pl. 12, fig. 5), USNM 689384 (unit 75, sec. 2; Pl. 12, fig. 6).

Rotalipora greenhornensis (Morrow)
(Pl. 11, figs. 7, 8)

Globorotalia greenhornensis Morrow, 1934, p. 199, Pl. 39, fig. 1. For synonymy, see Eicher, 1972, p. 186.

Material. Illustrated specimens are from units 61 and 75, section 2. Several specimens occur in most samples from the Greenhorn Formation; preservation throughout is good, except for excessive amounts of secondarily added carbonate and corrosion of certain specimens. The greatest diameter of the specimen in Plate 11, figure 7, is 0.43 mm; that in figure 8 is 0.51 mm, which represents the approximate maximum reached by this species in the Greenhorn Formation in Cimarron County.

Diagnosis. A large, trochospiral discoid form with periphery keeled, moderately lobate to almost entire; last whorl with 7 to 9 overlapping chambers; supplementary apertures umbilical or only slightly extraumbilical.

Remarks. Pessagno (1967, p. 289-292; 295-297) defined precisely the differences between *Rotalipora greenhornensis* and *R. appenninica* (O. Renz). Only a few specimens from Cimarron County have sufficiently pronounced convexity on the evolute side to fall into the range of *R. appenninica* as defined by Pessagno (1967). However, numerous transitional forms occur, along with an abundance of typical *R. greenhornensis*. Because of this we have not recognized *R. appenninica* in our samples.

Plate 11, figure 7, shows a type of variation noted in some specimens of *Rotalipora greenhornensis* within the upper part of its range in the Western Interior (Hartland Member in Cimarron County), where the final two or three chambers have assumed a shape similar to that of the last chambers in *R. cushmani*. A similar, but more globular, variation was illustrated by Eicher and Worstell (1970, Pl. 12, figs. 3a and 3b).

Stratigraphic and Geographic Position. This widely recognized species is recorded from sections 2 and 3 in Cimarron County. The oldest unit that contains *Rotalipora greenhornensis* is unit 17, the uppermost unit of the Graneros Shale at section 2. The species is rare in the Lincoln Member of the Greenhorn Formation through unit 34, section 2, and gradually but discontinuously increases in abundance upward through the Hartland Member until reaching a maximum in unit 75, section 2.

The first appearance of *Rotalipora greenhornensis* considerably below that of *R. cushmani* seems to be typical throughout the geographic range of the two species.

Figured Specimens. Well-preserved individual, USNM 689385 (unit 61, sec. 2; Pl. 11, fig. 7); second specimen, USNM 389386 (unit 75, sec. 2; Pl. 11, fig. 8).

Genus *Anaticinella* Eicher, 1972
Anaticinella multiloculata (Morrow)
(Pl. 12, fig. 8)

Globorotalia? multiloculata Morrow, 1934, p. 200, Pl. 31, figs. 3, 5. For synonymy, see Eicher, 1972, p. 185.

Material. Only a few poorly preserved specimens are available, all from unit 75, section 2. The figured individual (Pl. 12, fig. 8) is the largest of the lot and has a maximum diameter of 0.50 mm.

Diagnosis. A large species containing two to three whorls, with 6 to 9 (usually 7 or 8) globular chambers in the last whorl; periphery lobate; umbilicus wide and craterlike, and contains regular, lipped supplementary apertures.

Remarks. Eicher (1972, p. 184-190, Pls. 1, 2) has adequately discussed this species and has shown conclusively that the genus *Anaticinella* Eicher is a derivative of *Rotalipora greenhornensis* (Morrow).

Stratigraphic and Geographic Position. This species was recovered only from unit 75 of section 2. This occurrence is in the uppermost part of the Hartland Member, Greenhorn Formation. *Anaticinella multioculata* is known to occur at levels lower in Hartland equivalents in the Gulf Coast area, corresponding generally to the lowest occurrence of the species in the Western Interior (Eicher and Worstell, 1970, p. 314). The upper limit of occurrence of *A. multiloculata* at this locality is not known because of deep weathering of the lower beds in the Bridge Creek Limestone Member, which has eliminated recoverable microfossils. Eicher (1972, p. 184, 185) indicated a short range (latest Cenomanian) for this species. In the Eagle Ford Formation of north Texas, *A. multiloculata* occurs in rocks like the Hartland Member at the same biostratigraphic level, just beneath the *Sciponoceras gracile* Zone, and ranges up toward the base of the *Mytiloides mytiloides* Zone. This is the level (Upper Cenomanian) at which the earliest double-keeled planktonic foraminifer, *Marginotruncana difformis* (Gandolfi), occurs in the Eagle Ford Formation in Texas.

Figured Specimen. A moderately well preserved specimen, USNM 689387 (unit 75, sec. 2; Pl. 12, fig. 8).

Family MARGINOTRUNCANIDAE Pessagno, 1967

Genus *Whiteinella* Pessagno, 1967
Whiteinella aprica (Loeblich and Tappan)
(Pl. 12, figs. 10, 11)

Ticinella aprica Loeblich and Tappan, 1961, p. 292, Pl. 4, figs. 14–16. For synonymy, see Eicher and Worstell, 1970, p. 314.

Material. This species is present in considerable numbers in samples beginning with unit 75, Hartland Member of the Greenhorn Formation, and upward through unit 99 of section 2. Preservation is uniformly poor in the Bridge Creek Limestone Member (units 93, 97, and 99), but the sample from unit 75 (uppermost part of the Hartland Member) contains several well-preserved individuals. The specimens illustrated (Pl. 12, figs. 10, 11) are from the latter sample. These two specimens have greatest diameters of 0.67 and 0.44 mm, respectively; other specimens range up to 0.70 mm in diameter.

Diagnosis. A large trochoid species with two to three whorls, containing 5 to 7 globular chambers in the last whorl; periphery moderately to strongly lobate and rounded; increase in chamber size with age is gradual; umbilicus wide, commonly covered by tegilla.

Remarks. This species is best distinguished from its close relative, *Hedbergella portsdownensis* (Williams-Mitchell), by the tegilla that cover the greater part of the umbilicus of *Whiteinella aprica*. However, when preservation is poor and the tegilla are broken away, the most useful distinguishing feature is the high trochoid spire of *H. portsdownensis*. The spiral side of *W. aprica* is commonly relatively flat, except where the last whorl is strongly inclined toward the umbilicus. Further discussion is found in Eicher and Worstell (1970, p. 314).

Stratigraphic and Geographic Position. *Whiteinella aprica* occurs in the upper part of the Hartland Member (unit 75, sec. 2) and ranges upward throughout the exposed part of the Bridge Creek Member. The form is common in unit 75 (Hartland) and in units 93, 97, and 99 (Bridge Creek), section 2. This species occurs elsewhere in the Greenhorn Formation of the Western Interior and in the Eagle Ford Formation in north Texas. In Texas *W. aprica* occurs a little below and within the zone of *Sciponoceras gracile*, but has not been found as high as the zone of *Mytiloides mytiloides*.

Figured Specimens. USNM 689388 (unit 75, sec. 2; Pl. 12, fig. 10), USNM 689389 (unit 75, sec. 2; Pl. 12, fig. 11).

Genus *Marginotruncana* Hofker, 1956
***Marginotruncana difformis* (Gandolfi)**
(Pl. 12, figs. 1–4, 9)

Globotruncana intermedia difformis Gandolfi, 1955, p. 49, Pl. 3, figs. 4a–5c.
Globotruncana marginata (Reuss). Eicher, 1966, Pl. 6, figs. 5, 6.
Praeglobotruncana renzi (Thalmann). Eicher (1966), Pl. 6, fig. 9.
Marginotruncana marginata (Reuss). Pessagno, 1967, Pl. 54, figs. 7–12, 16–18; Pl. 56, figs. 10–12; Pl. 69, figs. 5–7. (See also footnote 6, Pessagno, 1967, p. 307.)
Praeglobotruncana difformis (Gandolfi). Eicher and Worstell, 1970, Pl. 11, figs. 4a–4c, 5, 6a–6c.

Material. Numerous moderately well preserved, complete specimens occur within the Bridge Creek Limestone Member, Greenhorn Formation. All figured specimens are from unit 97, section 2. Greatest diameters of specimens in Plate 12 are as follows: 0.51 mm (fig. 1), 0.44 mm (fig. 2), 0.43 mm (fig. 3), 0.47 mm (fig. 4), and 0.63 mm (fig. 9). Other individuals of this species reach diameters of as much as 0.64 mm.

Diagnosis. A large trochoid species containing 2 to 2.5 whorls, with 4 to 6 chambers in the last whorl; periphery commonly narrowly truncate and double-keeled, at least on last whorl; last few chambers may be keelless, with a narrowly rounded periphery; umbilicus wide and shallow.

Remarks. Pessagno's (1967, p. 299–301) resurrection and emended description of *Marginotruncana* Hofker does not completely clarify the systematic position of that genus. However, we agree with Pessagno that there is room for the generic concept of *Marginotruncana* between *Praeglobotruncana* and *Globotruncana* and that forms referred to *Marginotruncana* represent an important step in the evolution of the globotruncanid stock. In addition, the first appearance of truly double-keeled globigerinaceans (usually *M. difformis*) is of significant biostratigraphic value all over the world and is a recognizable, important evolutionary event.

Stratigraphic and Geographic Position. *Marginotruncana difformis* occurs in section 2 from unit 93 to the top of the exposure (unit 102). Because of lack of recovery of specimens in the weathered exposures of units 76 through 92, the lower limit of occurrence of this important species is not known. Elsewhere in the Western Interior its lower limit occurs close to the base of the Bridge Creek Limestone Member of the Greenhorn Formation (Eicher and Worstell, 1970). In Texas, *M. difformis* occurs at and slightly above the top of the *Sciponoceras gracile* Zone and overlaps the upper range of the youngest *Rotalipora* species.

Figured Specimens. Plate 12: USNM 689390 (fig. 1), USNM 689391 (fig. 2), USNM 689392 (fig. 3), USNM 689393 (Fig. 4), USNM 689394 (fig. 9). All are from unit 97, section 2.

PLATE SECTION

1
2
3
4
5
6

PLATE 1

Figure 1. *Inoceramus pictus pictus* Sowerby. Plastoholotype (×1), adult left valve of British Museum no. 43272, Chalk Marl, Guildford, England, illustrated for comparison of ornament with American species of this lineage; USNM 169393.

Figure 2. *Inoceramus ginterensis* Pergament. Hypotype (×2), internal mold, juvenile right valve showing abnormally close, regular juvenile rugae; typical growth lines (see Pl. 3, figs. 2, 4, 6) not preserved; USNM 169352. Unit 4, section 3, lower part of Hartland Member.

Figure 3. *Inoceramus pictus gracilistriatus* Kauffman and Powell, n. subsp. Holotype (×1), composite internal mold, adult left valve with typical ornament and form; USNM 169394. Unit 4, section 3, lower part of Hartland Member.

Figure 4. *Inoceramus flavus pictoides* Sornay. Hypotype (×1), composite mold, adult right valve, typical of subspecies; USNM 169395. Unit 3, section 3, lower part of Hartland Member.

Figures 5, 6. *Inoceramus prefragilis prefragilis* Stephenson. Paratypes (×1) from Lewisville Member, Woodbine Formation, Grayson County, Texas, zone of *Acanthoceras wintoni*, Middle Cenomanian, illustrated for comparison with younger Oklahoma species within lineage. 5, Internal mold, moderate-sized adult right valve, with abnormally weak rugae and posteroventrally trending sulcus; USNM 169314, 6, Internal mold, typical small adult left valve; USNM 169315. Note coarse regular growth lines, coarse rugae, and prominent though shallow sulcus.

1
2
3
4
5
6
7
8
9
10

PLATE 2

Figures 1, 3, 6, 8, 10. *Inoceramus prefragilis stephensoni* Kauffman and Powell, n. subsp. Unit 22, section 2, lower part of Lincoln Member, regional zone of *Dunveganoceras pondi*. 1, Paratype (×2), internal mold, umbonal area, left valve, anterior view, showing weakly developed lunule, regular, moderately weak juvenile growth lines; USNM 169396. 3, Paratype (×1), anterior view, umbo of right valve, latex cast of composite mold showing small lunule, lower convexity than left valve (fig. 1); USNM 169397b. 6, Paratype (×1), composite mold, right valve, posterior auricle missing, showing a typical, slight flattening of posteroventral growth line trace, as in *I. prefragilis prefragilis* (fig. 9), but without development of sulcus; USNM 169327. 8, Paratype (×1), latex cast of external mold, right valve with growth lines slightly finer than normal; USNM 169326b. 10, Holotype (×1), composite(?) mold, right valve with characteristic growth line trace and development, typically lacking strong rugae; USNM 169325.

Figure 2. *Inoceramus* sp. aff. *I. tenuis* Mantell (s. l.). Figured specimen, latex cast of external(?) mold (×1), umbonal area of left valve, beak broken, showing fine even concentric lamellae; USNM 169398b. Unit 24, section 2, lower part of Lincoln Member, regional zone of *Dunveganoceras pondi*.

Figure 4. *Inoceramus flavus pictoides* Sornay. Hypotype (×1), crushed incomplete internal mold, left valve showing typical coarse regular growth lines, prominent rugae; USNM 169400. Unit 3, section 3, lower part of Hartland Member, regional zone of *Dunveganoceras albertense*.

Figure 5. *Inoceramus tenuistriatus*? Nagao and Matsumoto. Figured specimen (×1), composite mold of incomplete left valve showing fine growth lines, regularly alternating rugae, prosogyrous beak-umbo typical of species; USNM 169399. Unit 81, section 2, upper part of lower Bridge Creek Limestone Member, zone of *Sciponoceras gracile*.

Figures 7, 9. *Inoceramus prefragilis prefragilis* Stephenson. Paratypes (×1), from the Lewisville Member, Woodbine Formation, zone of *Acanthoceras wintoni*, Grayson County, Texas, illustrated for comparison with younger species of lineage. 7, Internal mold with shell adhering, small adult left valve, with typical ornament and erect form, weak early sulcus; USNM 169317. 9, Internal mold with shell, left valve with unusually prominent sulci and rugae, characteristic change from adult to gerontic ornament; USNM 169316.

1
2
3
4
5
6
7
8
9
10

PLATE 3

Figures 1–4, 6. *Inoceramus ginterensis* Pergament. Hypotypes (figs. 1 and 2, ×2; figs. 3, 4, and 6, ×1), regional zone of *Dunveganoceras albertense*, lower (figs 1, 2, 4, 6) and middle (fig. 3) part of Hartland Member. 1, A pair of slightly displaced juvenile valves with typical fine, equal, early ornament, demonstrating low energy of the environment; USNM 169359. Unit 4, section 3. 2, Composite mold, small adult right valve, showing typical juvenile ornament of fine equal growth lines, sharp break in juvenile to adult ornamentation; USNM 169351. Unit 4, section 3. 3, Internal mold, right valve, medium-sized adult right valve; USNM 169350. Unit 24, section 3. 4, Composite mold, typical medium-sized adult left valve with juvenile and adult ornament preserved; USNM 169353. Unit 4, section 3. 6, Composite mold, left valve, typical ornament, medium-sized adult; USNM 169358. Unit 3, section 3.

Figures 5, 8. *Inoceramus* sp. aff. *I. tenuis* (s. l.) Mantell. Figured specimens, composite molds. Lower part of Lincoln Member, regional zone of *Dunveganoceras pondi*. 5, Juvenile left valve (×2) showing typical ornament of close-spaced, flat-lying, equal lamellae; USNM 169404b. Unit 24, section 2. 8, Adult right valve (×1) with ornament characteristic of *I. tenuis* but more quadrate, erect in form and with low beak-umbo; USNM 169405. Unit 22, section 2.

Figures 7, 9, 10. *Inoceramus tenuistriatus*? Nagao and Matsumoto. Figured specimens (×2), composite molds from the Greenhorn Formation. 7, Juvenile left valve, showing early ornament of regular raised growth lines, and first rugae separating juvenile and adult stages, USNM 169401. Unit 62, section 2, upper part of Hartland Member, regional zone of *Dunveganoceras albertense*. 9, Right valve of small adult with typical quadrate outline and regularly alternating weak rugae and fine growth lamellae; USNM 169402. Unit 81, section 2, upper part of the lower Bridge Creek Limestone Member, zone of *Sciponoceras gracile*. 10, Laterally compressed small adult right valve, auricle missing, ornament typically developed; USNM 169403. Unit 58, section 2, middle part Hartland Member, regional zone of *Dunveganoceras albertense*.

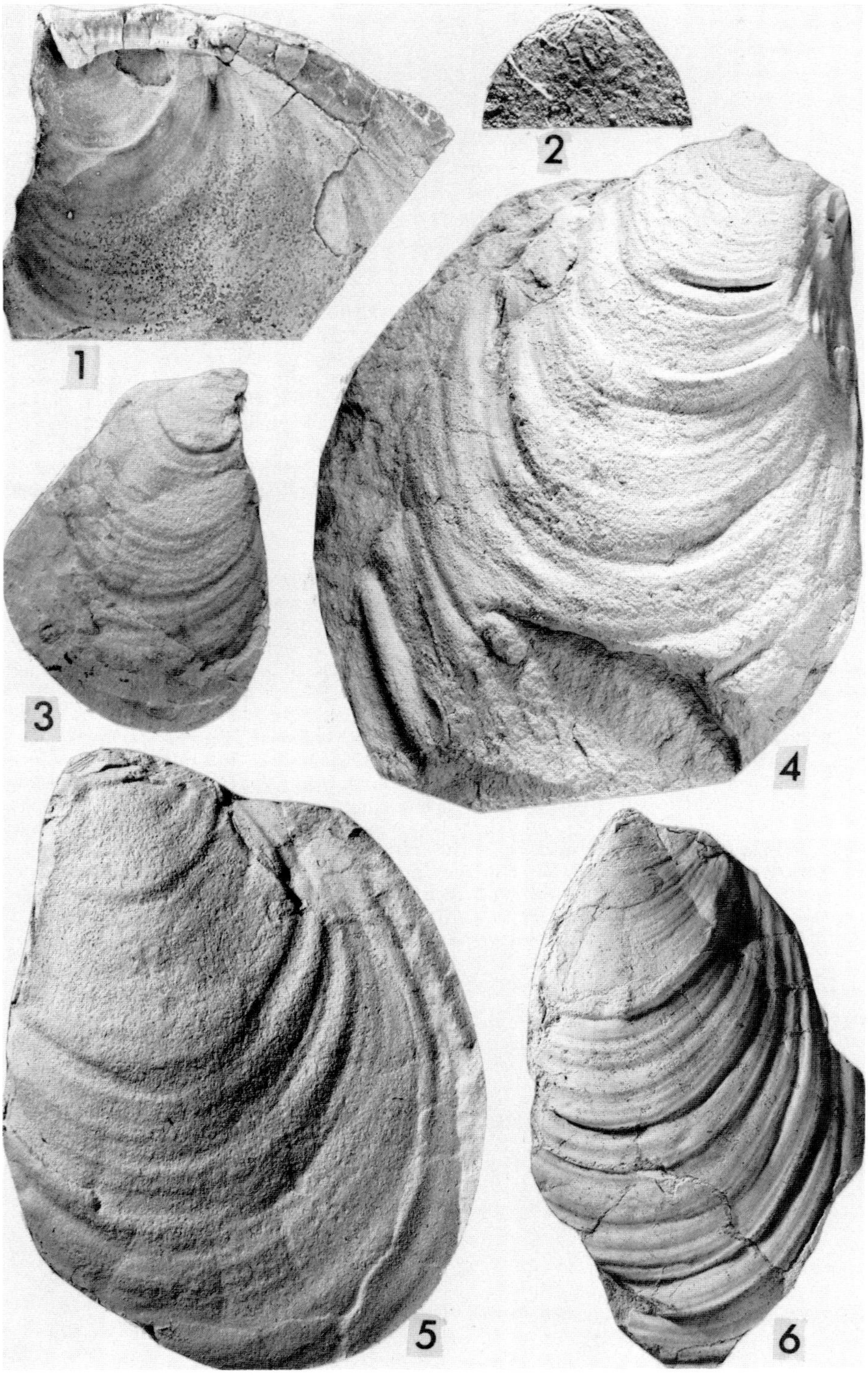
1
2
3
4
5
6

PLATE 4

Figures 1, 3–6. *Inoceramus ginterensis* Pergament. Hypotypes (×1). Unit 4, section 3, lower part of Hartland Member, regional zone of *Dunveganoceras albertense*. 1, Interior view, right valve, showing part of ligamental plate, nature of resilifers, umbonal septum (anterior part missing), and well-marked posterior part of entire pallial line; USNM 169347. 3, Internal mold, right valve, small adult, with typical prosogyrous beak and early development of rugae; USNM 169349. 4, Composite mold, right valve, typical large adult showing distribution of rugae, concave anterior margin, projecting umbo with faint juvenile ornament; USNM 169348. 5, Internal mold, typical adult left valve, with well-developed posterior auricle, rugae; USNM 169355. 6, Composite mold, secondarily laterally compressed left valve with typical, well-defined, adult ornament; USNM 169354.

Figure 2. *Entolium* (*Entolium*) *gregarium* Kauffman and Powell, n. sp. Paratype (×2), internal mold of small adult valve showing dorsally casts of crura and lateral resilifer extensions typical of genus; USNM 169338b. Note slight upward inclination of dorsal auricle edge, rare on this species but typical for genus and subgenus; other auricle broken. Unit 3, section 3, lower part of Hartland Member, regional zone of *Dunveganoceras albertense*.

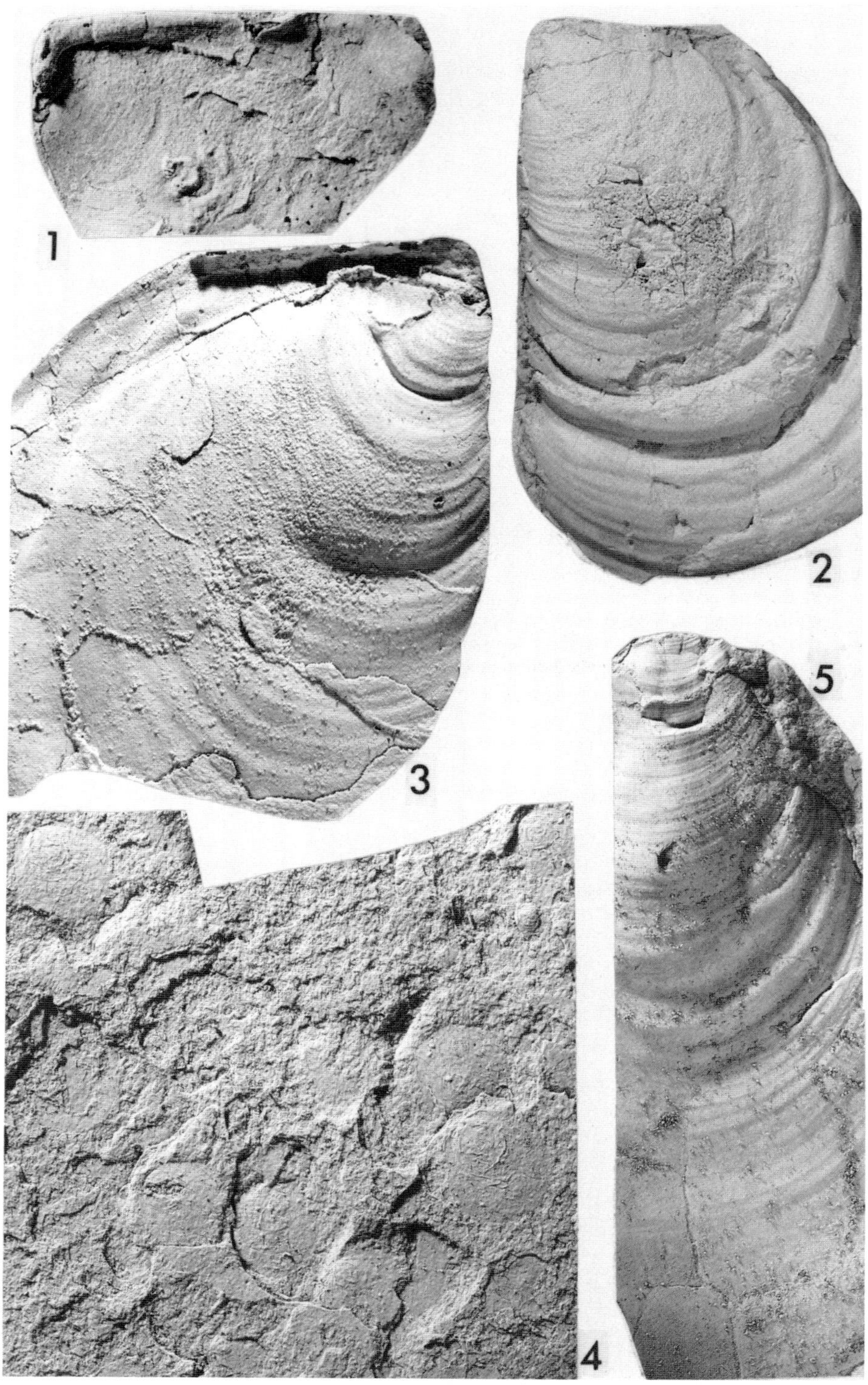
1
2
3
4
5

PLATE 5

Figure 1. *Mytiloides*? sp. aff. *M.*? *crippsi* Mantell. Figured specimen (×1), latex cast of internal mold showing rounded form, weak ornament, thickened ligamental plate, very small umbonal septum, and very small internal rib (longitudinally fractured) separating auricle from disc on shell interior; USNM 172868b. Unit 10, section 2, lower part of upper shale member, Graneros Shale.

Figures 2, 3, 5. *Inoceramus ginterensis* Pergament. Hypotypes (×1), composite molds. Unit 4, section 3, lower part of Hartland Member, regional zone of *Dunveganoceras albertense*. 2, Medium-sized adult left valve with typical ornament, unusually depressed beak-umbo; USNM 169356. 3, Latex mold of shell interior showing posterior part of pallial line, abnormally weak rugae; USNM 169347. 5, Large left valve, anterior and posterior flanks cut away, with very well preserved fine ornament of irregular lamellae; USNM 169357.

Figure 4. *Entolium* (*Entolium*) *gregarium* Kauffman and Powell, n. sp. Holotype (middle right) and paratypes, a slab (×1) containing several single and paired adult valves on one bedding plane, presumably a mass mortality; USNM 169338a–169338g. Unit 3, section 3, lower part of Hartland Member, regional zone of *Dunveganoceras albertense*.

1
2
3
4
5
6
7
8
9
10
11
12
13
14
15
16

PLATE 6

Figures 1, 2. *Discinisca*? sp. Figured specimen, uncoated (fig. 1, ×2) and coated (fig. 2, ×1) dorsal views of a typical specimen, brachial valve, showing lamellate ornament on thick shell, flattened posterior flank of apex; USNM 169309. Unit 22, section 2, Lincoln Member, regional zone of *Dunveganoceras pondi*.

Figures 3, 6. *Mytiloides opalensis* (Böse). Hypotypes. Unit 90, section 2, lower middle part of the Bridge Creek Limestone Member, local zone of *M. opalensis*. 3, Composite mold, small adult right valve (×1), showing typical form and ornament of regular rugae without strong intercalated growth lines; USNM 169310a. 6, Internal mold, juvenile part of small adult right valve (×1) with characteristic coarse early rugae; USNM 169310b.

Figure 4. *Mytiloides* sp. transitional between *M. opalensis* (Böse) and *M. mytiloides* (Mantell). Composite mold (×1), small adult right valve with shape of *M. opalensis* and ornament plan of *M. mytiloides*; rugae and growth lines conspicuously finer than on *M. mytiloides*; USNM 169311. Unit 86, section 2, Bridge Creek Member.

Figure 5. *Mytiloides* sp. aff. *M. duplicostatus* (Anderson). Figured specimen (×1), composite mold, left valve, beak broken, rugae showing trend toward distinct pairing; USNM 169312. Unit 90, section 2, Bridge Creek Limestone Member.

Figures 7, 10. *Mytiloides* sp. aff. *M. submytiloides* (Seitz). Figured specimens. Unit 62, section 2, upper part of Hartland Member, regional zone of *Dunveganoceras albertense*. 7, Internal mold (×1), medium-sized adult left valve with characteristic weak rugae and irregular growth lines; USNM 169334. 10, Internal mold (×2), small adult right valve with regular small rugae on juvenile shell; irregular, primarily weak adult rugae, shallow posteroventrally trending sulcus behind umbonal fold; USNM 169335.

Figures 8, 9. *Entolium* (*Entolium*) *gregarium* Kauffman and Powell, n. sp. Selected paratypes (×2) from a single slab; USNM 169340. Unit 3, section 3, lower part of Hartland Member, regional zone of *Dunveganoceras albertense*. 8, Composite molds, two adult valves with well-preserved auricles typically sloping slightly downward from beak and traces of crura; USNM 169340a (left), 169340b (right). 9, An unusually inclined right(?) valve with elongated posteroventral margin, crura partially preserved; USNM 169340c.

Figures 11-16. *Mytiloides mytiloides* (Mantell). Hypotypes (×1), composite molds. Unit 100, section 2, lower part of the upper Bridge Creek Limestone Member, local *M. mytiloides* Zone. 11, Juvenile right valve with small closely spaced rugae, typical early growth characters; USNM 169387. 12, Juvenile left valve with typically prominent beak, weak rugae and fine crowded growth lines; USNM 169386a (latex cast). 13, Small adult left valve with characteristic ornament and form; USNM 169385. 14, Small adult right valve with abnormally weak ornament, well-developed auricle; USNM 169384. 15, Juvenile left valve, unusually erect (transitional to *M. latus*), with typical early ornament of close equal rugae and fine growth lines; USNM 169386b (latex composite mold, same rock as fig. 12). 16, Medium-sized adult left valve, beak and auricle missing, with very characteristic adult ornamentation of regularly alternating rugae and raised growth lines crossing rugae at angle; USNM 169383.

1
2
3
4
5
6

PLATE 7

Figure 1. ***Mytiloides opalensis*** **(Böse). Typical adult right valve (×1), copied from Plate 39, figure 3b of Seitz (1934 [1935]; as *Inoceramus labiatus* var. *opalensis* Böse) from "Ziegenberg bei Wolfsdorf," Lower Turonian (compare with Pl. 6, figs. 3, 6).**

Figures 2, 6. ***Mytiloides*** **sp. aff. *M. duplicostatus* (Anderson). Figured specimens (×1), fragments of right (fig. 2) and left (fig. 6) valves showing characteristic ornament of paired coarse raised growth lines on major rugae and small number of regular (early adult) to irregular raised growth lines intercalated between rugae. Unit 90, section 2, Bridge Creek Limestone Member. 2, USNM 169388. b, USNM 169392.**

Figure 3. ***Mytiloides*** **sp. transitional between *M. opalensis* (Böse) and *M. mytiloides* (Mantell). Figured specimen (×1), the late erect, finely and evenly ornamented form (see Fig. 4 and compare with Pl. 6, fig. 4); right-lateral view of small adult; USNM 193495. Unit 96, section 2.**

Figure 4. ***Inoceramus ginterensis*** **Pergament. Typical medium-sized adult left valve (×1), copied from Pergament (1966, Pl. 27, fig. 1a), Cenomanian of the Pacific coast, USSR (compare with Pl. 3, figs. 2–4, 6; Pl. 4, figs. 1, 3–6; Pl. 5, figs. 2, 3 [especially], 5).**

Figure 5. ***Mytiloides labiatus labiatus*** **(Schlotheim). Left-lateral view (×1.1) of an adult valve, somewhat crushed, considered by Seitz (1934 [1935]; 1971 oral commun.) to be closest to Schlotheim's lost type specimen. Copied from Seitz (1934 [1935], Pl. 38, fig. 1). Compare with previous American concept of species (*M. mytiloides*, Pl. 6, figs. 11–16).**

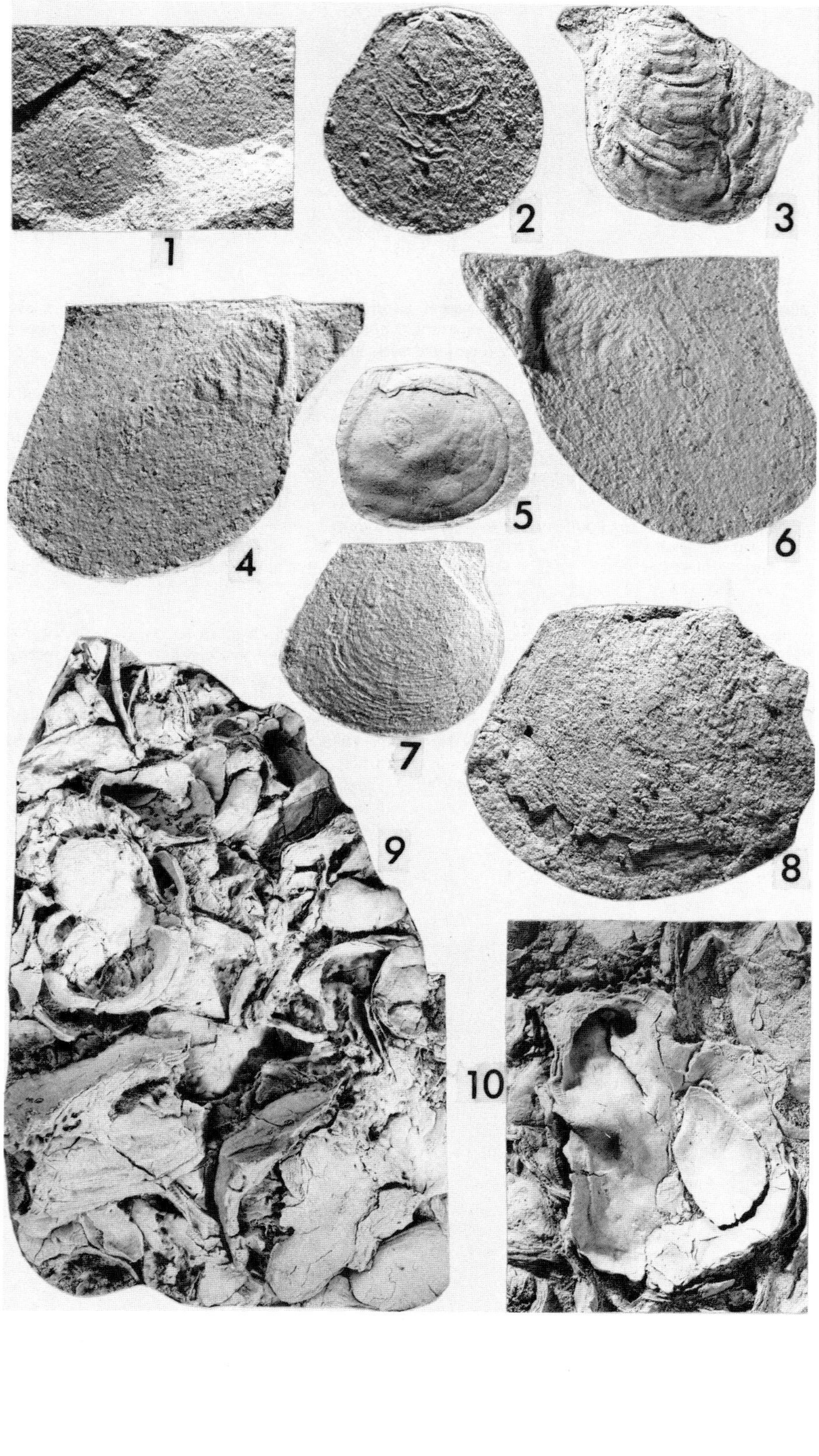
1
2
3
4
5
6
7
8
9
10

PLATE 8

Figures 1, 2. *Entolium (Entolium) gregarium* Kauffman and Powell, n. sp. Composite molds. Unit 3, section 3, lower part of Hartland Member, regional zone of *Dunveganoceras albertense*. 1, Paratypes (×2), small adult valves with typical form, auricles sloping gently away from beak, traces of crura preserved; USNM 169339a (left), 169339b (right). 2, Holotype (×3), complete typical small adult valve with traces of diagonal crura on either side of umbo, of small subtriangular dorsocentral resilifer and short lateral grooves adjacent to it; USNM 169338a. Part of a large slab with many specimens on a single bedding plane.

Figure 3. *Phelopteria* sp. A. Figured specimen (×2), latex cast of a composite mold, left valve, with posterior auricle partly broken, anterior auricle typically developed, unusually coarse rugae for genus; USNM 169406b. Unit 62, section 2, upper part of Hartland Member, regional zone of *Dunveganoceras albertense*.

Figures 4, 6. *Phelopteria minuta* Kauffman and Powell, n. sp. Holotype (×2). Unit 24, section 3, middle part of Hartland Member, regional zone of *Dunveganoceras albertense*. 4, Latex composite mold (of USNM 169320a, fig. 6) of right valve showing fine crowded growth lines and their sinuous trace; USNM 169320b. 6, Interior view, complete, typical right valve with fully developed auricles; USNM 169320a.

Figure 5. *Pycnodonte newberryi* (Stanton). Internal mold (×2), right valve with parts of adhering shell, showing well-defined pallial line (a raised rim impressed into the mold), typical circular posterocentral adductor muscle insertion area, and sinuous elongate chomata on auricles; USNM 169407. Unit 100, section 2, upper part of Bridge Creek Limestone Member, zone of *Mytiloides mytiloides* (Mantell).

Figures 7, 8. *Phelopteria* quadrate species aff. *P. minuta* Kauffman and Powell. Figured specimens, small adult right valves with anterior auricles broken off. 7, Composite internal mold (×2), most complete specimen with part of auricle preserved anteriorly, typical ornament and lack of posterior extension on posterior auricle (compare with *P. minuta*, figs. 4, 6); USNM 169321. Note erect quadrate form, relatively greater length than in *P. minuta*. Unit 24, section 3, middle part of Hartland Member, regional zone of *Dunveganoceras albertense*. 8, Outer shell surface (×3), with recrystallized shell intact showing typical form, posterior auricle, and fine equally developed growth lines; USNM 169322. Unit 62, section 2, upper part of Hartland Member, regional zone of *D. albertense*.

Figures 9, 10. *Ostrea beloiti* Logan. Hypotypes (×1). Unit 14 (*O. beloiti* biostrome), section 2, upper part of Graneros Shale, regional zone of *Acanthoceras amphibolum*. 9, Part of a large slab showing typical *O. beloiti* biostromal fabric, irregular cupped form of attached left valves (lower left), smooth, flat, slightly curved right valves (middle upper right, left); USNM 169319. 10, Moderately large left (attached) valve with small, independent right valve nested inside it (from a large biostromal slab); USNM 169318. Note fine, close-set, slightly sinuous chomata within commissure, posteroventral adductor insertion areas, shallow resilifers and deep subumbonal cavity.

1
2
3
4
5
6
7

PLATE 9

Figures 1, 3, 4. *Stomohamites simplex cimarronensis*, n. subsp. See also Figures 5 and 6. Unit 4, section 3, Hartland Member. 1, Holotype (×0.8), lateral view; USMN 167160. 3, Paratype (×0.75), lateral view; USNM 167164. 4, Paratype (×1.75), lateral view; USNM 167162.

Figure 5. *Hemiptychoceras* sp. Figured specimen (×1.6), lateral view; USNM 167161. See also Figure 7. Unit 4, section 3, Hartland Member.

Figures 2, 6. *Desmoceras*?, s. l., sp. Figured specimens. 2, Lateral view (×2); USNM 167169. Unit 4, section 3, Hartland Member. 6, Lateral view (×1.3); USNM 167165. Unit 13, section 3, Hartland Member.

Figure 7. *Pseudouhligella* sp. Figured specimen (×2), lateroventral view of rubber cast; USNM 167166. Unit 4, section 3, Hartland Member.

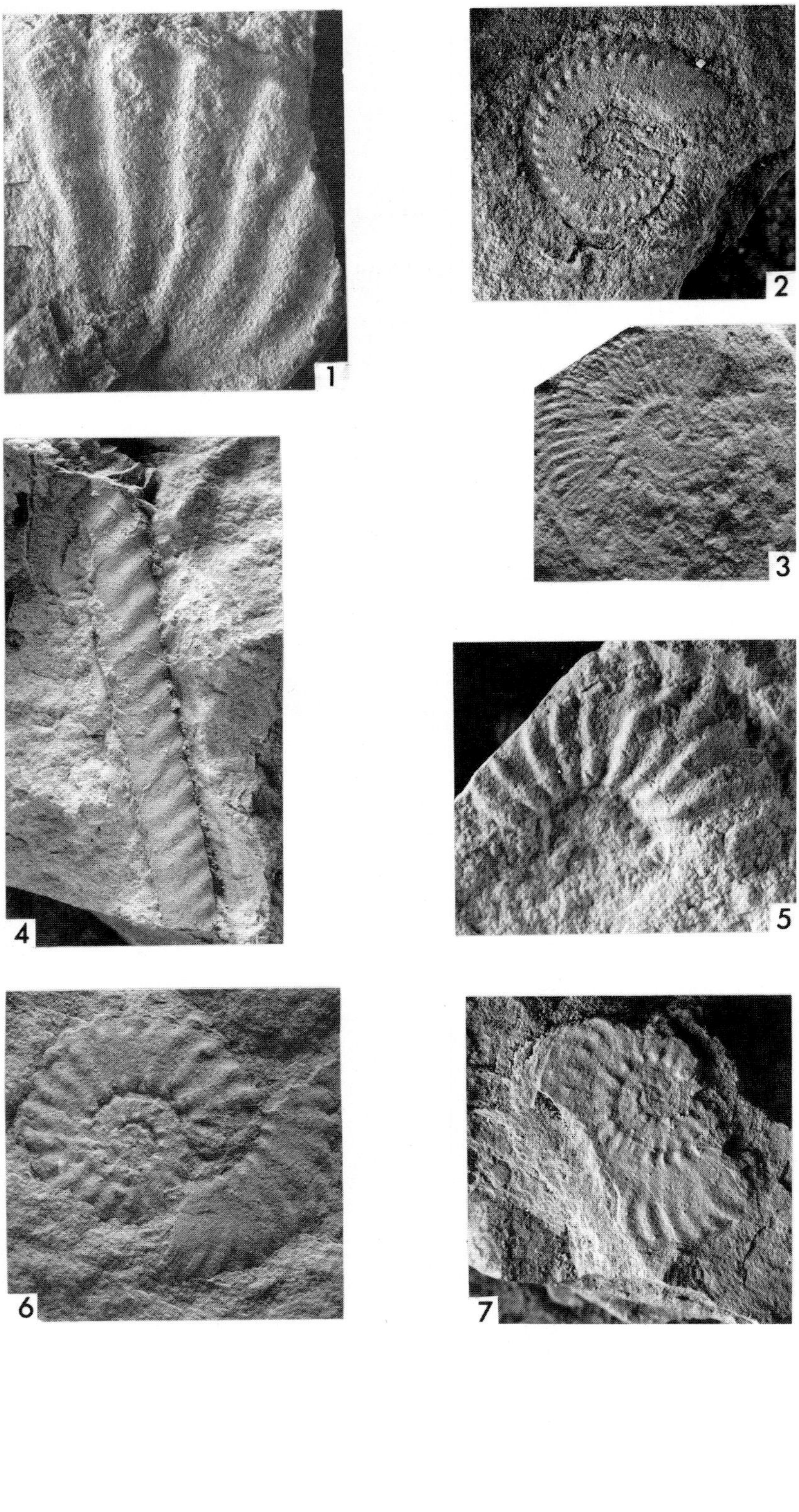

PLATE 10

Figure 1. *Pseudocalycoceras* sp. Figured specimen (×2), lateral view; USNM 167173. Unit 22, section 2, lower part of Lincoln Member.

Figures 2, 6, 7. *Acanthoceras* group "B." Figured specimens (×2). 2, Lateral view; USNM 167168. Unit 4, section 3, Hartland Member. 6, Lateral view; USNM 167167. Unit 4, section 3, Hartland Member. 7, Lateral view; USNM 167170. Unit 3, section 3, Hartland Member.

Figure 3. *Acanthoceras* group "A." Figured specimen (×2), lateral view of imprint; USNM 167174. Unit 2, section 3, Hartland Member.

Figure 4. *Sciponoceras gracile* (Shumard). Hypotype (×0.8), lateral view of phragmocone; USNM 167172. Unit 81, section 2, lower part of Bridge Creek Limestone Member.

Figure 5. *Quitmaniceras* sp. cf. *Q. brandi* Powell. Figured specimen (×1.5), lateral view; USNM 167171. Unit 86, section 2, Bridge Creek Limestone Member.

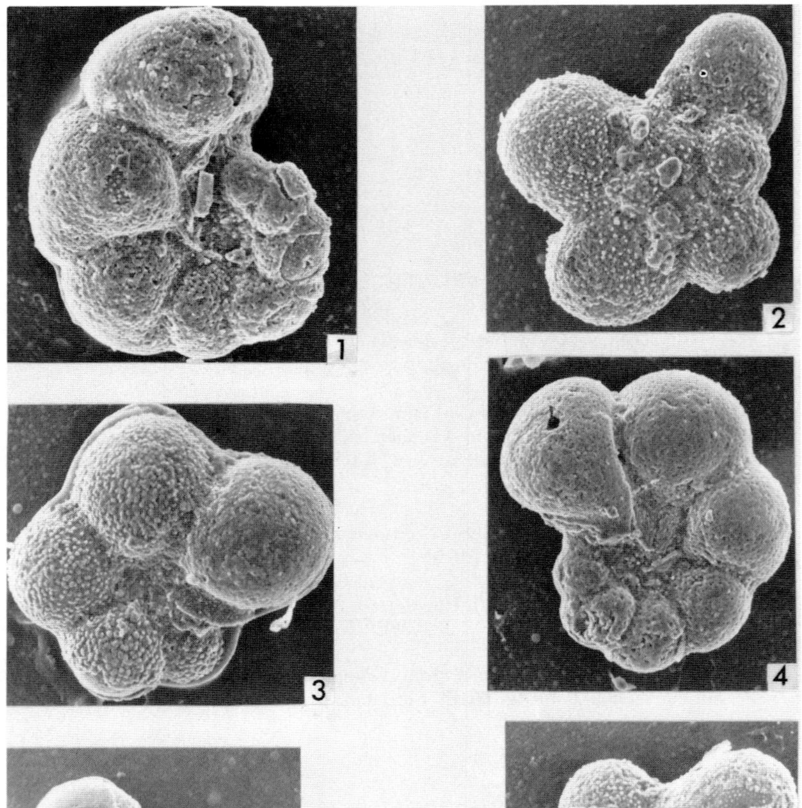
1
2
3
4
5
6
7
8

PLATE 11

All figures stereoscan

Figures 1, 4. ***Globigerinelloides bentonensis*** **(Morrow). Hypotypes (×120). Unit 44, section 2, Lincoln Member. 1, USNM 689376. 4, USNM 689377.**

Figure 2. ***Clavihedbergella simplex*** **(Morrow). Hypotype (×120); USNM 689382. Unit 44, section 2, Lincoln Member.**

Figure 3. ***Hedbergella delrioensis*** **(Carsey). Hypotype (×120); USNM 689379. Unit 34, section 2, Lincoln Member.**

Figure 5. ***Heterohelix globulosa*** **(Ehrenberg). Hypotype (×135); USNM 689375. Unit 44, section 2, Lincoln Member.**

Figure 6. ***Hedbergella amabilis*** **Loeblich and Tappan. Hypotype (×120); USNM 689378. Unit 44, section 2, Lincoln Member.**

Figures 7, 8. ***Rotalipora greenhornensis*** **(Morrow). Hypotypes (×90). 7, USNM 689385. Unit 61, section 2, Hartland Member. 8, USNM 689386. Unit 75, section 2, Hartland Member.**

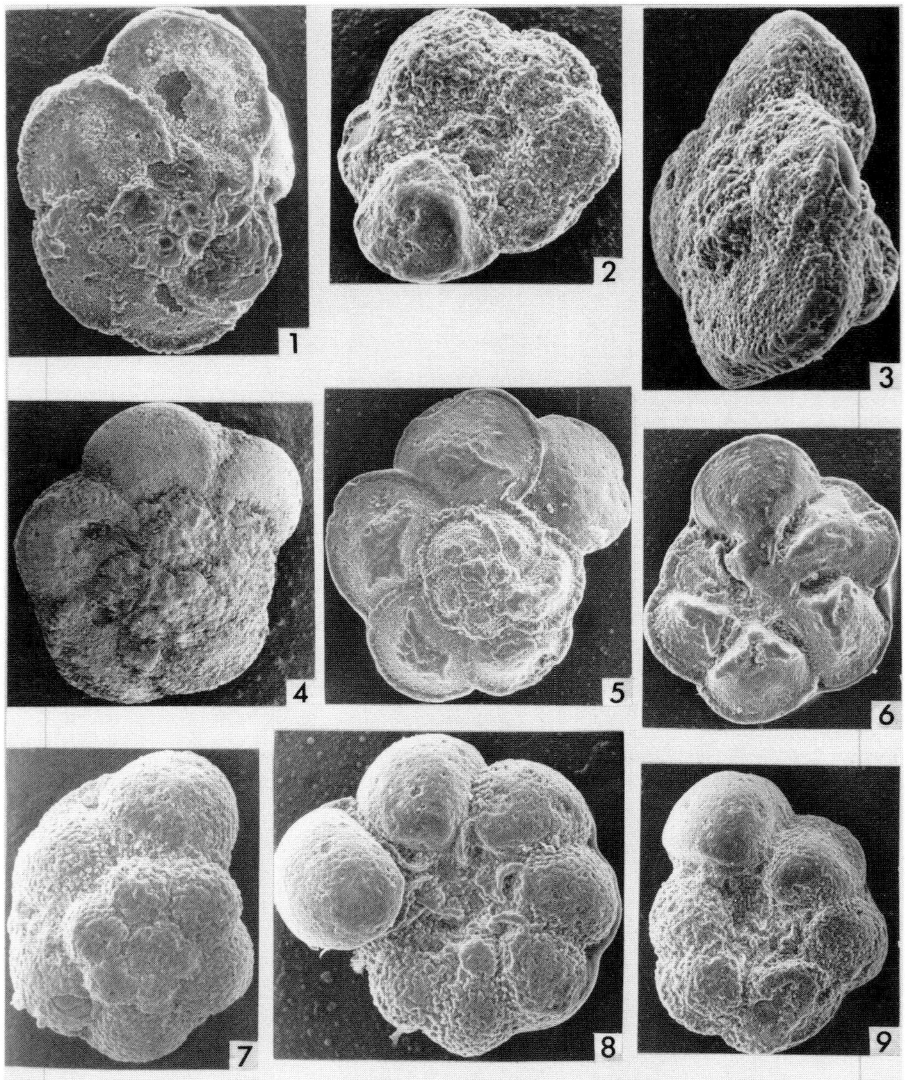
1
2
3
4
5
6
7
8
9
10
11
12

PLATE 12

All figures stereoscan

Figures 1–4, 9. ***Marginotruncana difformis*** **(Gandolfi). Hypotypes. Unit 97, section 2, Bridge Creek Limestone Member. 1, USNM 689390 (×90). 2, USNM 689391 (×90). 3, USNM 689392 (×120). 4, USNM 689393 (×90). 9, USNM 689394 (×60).**

Figures 5, 6. ***Rotalipora cushmani*** **(Morrow). Hypotypes (×60). Unit 75, section 2, uppermost part of Hartland Member. 5, USNM 689383. 6, USNM 689384.**

Figures 7, 12. ***Hedbergella portsdownensis*** **(Williams-Mitchell). Hypotypes (×90). Unit 93, section 2, Bridge Creek Limestone Member. 7, USNM 689380. 8, USNM 689381.**

Figure 8. ***Anaticinella multiloculata*** **(Morrow). Hypotype (×90); USNM 689387. Unit 75, section 2, uppermost part of Hartland Member.**

Figures 10, 11. ***Whiteinella aprica*** **(Loeblich and Tappan). Hypotypes. Unit 75, section 2, uppermost part of Hartland Member. 10, USNM 689388 (×60). 11, USNM 689389 (×90).**

References Cited

Adkins, W. S., 1928, Handbook of Texas Cretaceous fossils: Texas Univ. Bull. 2838, 385 p., 37 pls.

——1933, The Mesozoic Systems in Texas, *in* Sellards, E. H., Adkins, W. S., and Plummer, F. B., The geology of Texas; Vol. 1, Stratigraphy: Texas Univ. Bull. 3232, p. 239–518, 33 text-figs. (1932).

Anderson, F. M., 1958, Upper Cretaceous of the Pacific coast: Geol. Soc. America Mem. 71, 378 p., 75 pls., 3 text-figs., 5 tables.

Bandy, Orville, 1967, Cretaceous planktonic foraminiferal zonation: Micropaleontology, v. 13, p. 1–31.

Bass, N. W., 1926, Geologic investigations in western Kansas, with special reference to oil and gas possibilities, Pt. II, The geology of Hamilton County, Kansas: Kansas Geol. Survey Bull. 11, p. 53–83, 2 pls., 14 text-figs.

Bass, N. W., Straub, C. E., and Woodbury, H. O., 1947, Structure contour map of the surface rocks of the Model anticline, Las Animas County, Colorado: U.S. Geol. Survey Oil and Gas Inv. Prelim. Map 68, text with map, scale 1 in. = 1 mi.

Bergquist, H. R., 1944, Cretaceous of the Mesabi Iron Range, Minnesota: Jour. Paleontology, v. 18, p. 1–30, Pls. 1–11.

——1949, The Woodbine of Cooke, Grayson and Fannin counties, Texas: U.S. Geol. Survey Oil and Gas Inv. Prelim. Map 98, text with map.

Bertin, Léon, 1939, Géologie et paléontologie (5th ed.): Paris, Lib. Larousse, 504 p., plates, illus., map, diagrams.

Böhm, Johannes, 1909, Geologie and Paläontologie du subherzynican Kreidemulde: Preuss. Geol. Landesanst. Abh., N. F., no. 56, p. 39–58, Pls. 9–14.

Bolli, H. M., Loeblich, A. R., Jr., and Tappan, Helen, 1957, Planktonic foraminiferal families Hantkeninidae, Orbulinidae, Globorotaliidae, and Globotruncanidae: U.S. Natl. Mus. Bull., no. 215, p. 3–50, Pls. 1–11, Text-Figs. 1–9.

Böse, Emilio, 1910, Neue Beiträge zur Kenntnis der mexikanischen Kreide: Centralbl. Mineralogie, Geologie u. Paläontologie, no. 20, p. 652–662, 657.

——1913, Algunas faunas del Cretácico superior de Coahuila y regiones limítrofes: México Inst. Geol. Bol., no. 30, p. 56, Pl. 8.

——1923, Algunas faunas Cretácicas de Zacatecas, Durango y Guerrero: México Inst. Geol. Bol.: no. 42, p. 219, Pl. 19.

Brongniart, A., 1822, *in* Cuvier, G., Recherches sur les ossemens fossiles, Vol. 2, Pt. 2: Paris, G. Dufour et E. d'Ocagne, p. 333–609.

Bronn, H. G., 1853-1856, Paläontologie, *in* Bronn, H. G., and Roemer, F., 1851–1856, Lethea Geognostica, oder Abbildungen und Beschreibung der für die Gebirgs-Formationen bezeichnendsten Versteinerungen; Vol. 3, Pt. 6: Mollassen-Periode: Stuttgart, E. Schweizerhart, 1130 p. [a similar work bearing the same title was begun by F. Roemer in 1876, and carried on by F. Frech].

Carsey, D. O., 1926, Foraminifera of the Cretaceous of central Texas: Texas Univ. Bull., no. 2612, 56 p., 8 pls.

Cobban, W. A., 1951, Scaphitoid cephalopods of the Colorado Group: U.S. Geol. Survey Prof. Paper 239, 39 p., 21 pls., 4 text-figs.

——1952, A new Cretaceous ammonite genus from Wyoming and Utah: Jour. Paleontology, v. 26, p. 758-760, 1 pl.

——1953, Cenomanian ammonite fauna from the Mosby Sandstone of central Montana: U.S. Geol. Survey Prof. Paper 234-D, p. 45-55, pls. 6-12, Text-Figs. 1-3.

——1958a, Two new species of *Baculites* from the Western Interior region: Jour. Paleontology, v. 32, p. 660-665, pls. 90, 91, 1 text-fig.

——1958b, Late Cretaceous fossil zones of the Powder River Basin, Wyoming and Montana, *in* Wyoming Geol. Assoc. Guidebook, 13th Ann. Field Conf., 1958, Powder River Basin: p. 114-119, 2 text-figs.

——1961, The ammonite family Binneyitidae in the Western Interior of the United States: Jour. Paleontology, v. 35, p. 737-758, Pls. 87-89, 5 text-figs.

——1962a, *Baculites* from the lower part of the Pierre Shale and equivalent rocks in the Western Interior: Jour. Paleontology, v. 36, p. 704-718, Pls. 105-108, 1 text-fig.

——1962b, New *Baculites* from the Bearpaw Shale and equivalent rocks of the Western Interior: Jour. Paleontology, v. 36, p. 126-135, Pls. 25-28, 1 text-fig.

——1963, Occurrence of the Late Cretaceous ammonite *Hoplitoplacenticeras* in Wyoming: U.S. Geol. Survey Prof. Paper 475-C, art. 76, p. C60-C61, fig. 76.1.

——1964, The Late Cretaceous cephalopod *Haresiceras* Reeside and its possible origin: U.S. Geol. Survey Prof. Paper 454-I, 21 p., 3 pls., 7 text-figs.

——1969, The Late Cretaceous ammonites *Scaphites leei* and *Scaphites hippocrepis* (DeKay) in the Western Interior of the United States: U.S. Geol. Survey Prof. Paper 618, 29 p., 5 pls., 21 text-figs.

Cobban, W. A., and Reeside, J. B., Jr., 1952, Correlation of the Cretaceous formations of the Western Interior of the United States: Geol. Soc. America Bull., v. 63, p. 1011-1044, charts 10a, 10b.

Cobban, W. A., and Scott, G. R., 1964, Multinodose scaphitid cephalopods from the lower part of the Pierre Shale and equivalent rocks in the conterminous United States: U.S. Geol. Survey Prof. Paper 438-E, 13 p., 4 Pls., 5 text-figs.

——1972, Stratigraphy and ammonite fauna of the Graneros Shale and Greenhorn Limestone near Pueblo, Colorado: U.S. Geol. Survey Prof. Paper 645, 101 p., 5 tables, 41 pls., 52 text-figs.

Conrad, T. A., 1857, Descriptions of Cretaceous and Tertiary fossils, *in* Emory, W. H., Report on the United States and Mexico boundary survey . . .: U.S. 34th Cong., 1st sess., House Executive Doc. 135, v. 1, pt. 2, p. 141-174, pls. 1-21.

——1874 (1875), Descriptions of new genera and species of fossil shells of North Carolina—Synopsis of the Cretaceous Mollusca of North Carolina—Remarks on some genera of shells, *in* Kerr, W. C., Report of the Geological Survey of North Carolina, Vol. 1: Raleigh, 325 p., 1 pl., 1 map, text illus.

Cox, L. R., 1969, Family Inoceramidae Giebel, *in* Moore, R. C., ed., Treatise on invertebrate paleontology, Pt. N, Vol. 1, Mollusca 6, Bivalvia: Boulder, Colo., Geol. Soc. America (and Univ. Kansas Press), p. N314-N321, Figs. C46-C49.

Cragin, F. W., 1893, A contribution to the invertebrate paleontology of the Texas Cretaceous: Texas Geol. Survey, Ann. Rept. No. 4, pt. 2, p. 139-294.

Credner, C. H., 1897, Elemente der Geologie (2nd ed.): Leipzig, W. Engelmann, xviii + 797 p., illus.

Cushman, J. A., 1946, Upper Cretaceous Foraminifera of the Gulf coastal region of the United States and adjacent areas: U.S. Geol. Survey Prof. Paper 206, 241 p., 66 pls.

Dacqué, E., 1939, Die Fauna der Regensburg-Kelheimer Oberkreide (mit Ausschluss der Spongien und Bryozoen): Bayer. Akad. Wiss., Mat.-Naturw. Abt., Abh., N. F., no. 45, 218 p., 17 pls.

Darton, N. A., 1905, Preliminary report on the geology and underground water resources of the central Great Plains: U.S. Geol. Survey Prof. Paper 32, 433 p., 72 pls., illus., maps.

——1906, Geology and underground waters of the Arkansas Valley in eastern Colorado: U.S. Geol. Survey Prof. Paper 52, 90 p., 28 pls.

——1909, Geology and water resources of the northern portion of the Black Hills and adjoining regions, in South Dakota and Wyoming: U.S. Geol. Survey Prof. Paper 65, 105 p., 24 pls., illus., maps, diagrams.

DeFord, R. K., 1927, Areal geology of Cimarron County, Oklahoma: Am. Assoc. Petroleum Geologists Bull., v. 11, p. 753-755.

Dietrich, W. O., 1924, Zur Paläontologie und Stratigraphie der Kreide und des Tertiärs in der Ostkaribischen Kordillere Venezuelas: Centralbl. Mineralogie, Geologie, u. Paläontologie, no. 6, p. 181-187.

Dixon, Fredrick, and Jones, T. R., 1878, The geology of Sussex (2nd ed.): Brighton, W. J. Smith, Pub., 169 p., illus., pls., folding map.

Donovan, D. T., 1954, Upper Cretaceous fossils from Traill and Geographical Society Øer, East Greenland: Medd. Grønland, v. 72, no. 6, 33 p., 3 pls.

Douglas, R. G., 1969, Upper Cretaceous planktonic foraminifera in northern California; Pt. 1, Systematics: Micropaleontology, v. 15, p. 151-209, pls. 1-11, Text-Figs. 1-6.

Ehrenberg, C. G., 1840, Über die Bildung der Kreidefelsen und des Kreidemergels durch unsichtbare Organismen: Königl. Akad. Wiss. Berlin, Physik, Abh., p. 59-147, pls. 1-4.

Eicher, D. L., 1960, Stratigraphy and micropaleontology of the Thermopolis Shale: Yale Univ. Peabody Mus. Nat. History Bull. 15, 126 p., 6 pls., 12 text-figs.

——1965, Foraminifera and biostratigraphy of the Graneros Shale: Jour. Paleontology, v. 39, p. 875-909, pls. 103-106, 6 text-figs.

——1966, Foraminifera from the Cretaceous Carlile Shale of Colorado: Cushman Found. Foram. Research Contr., v. 17, pt. 1, p. 16-31, pls. 4-6, 2 text-figs.

——1972, Phylogeny of the late Cenomanian planktonic foraminifer *Anaticinella multiloculata* (Morrow): Jour. Foram. Research, v. 2, no. 4, p. 184-194, 2 pls., 4 text-figs.

Eicher, D. L., and Worstell, Paula, 1970, Cenomanian and Turonian Foraminifera from the Great Plains, United States: Micropaleontology, v. 16, p. 269-324, Pls. 1-13, 12 text-figs.

Folinsbee, R. E., Baadsgaard, H., and Cumming, G. L., 1963, Correlation of mid-Cretaceous bentonites by K-Ar dating [abs.]: Geol. Soc. America Spec. Paper 73, p. 84.

Fraas, Eberhard, 1910, Der Petrefaktensammler: Stuttgart, K. G. Lutz, Pub., 276 p., 72 pls., illus.

Franks, P. C., Coleman, G. L., Plummer, Norman, and Hamblin, W. K., 1959, Cross-stratification, Dakota Sandstone (Cretaceous), Ottawa County, Kansas: Kansas Geol. Survey Bull. 134, pt. 6, p. 223-238, 2 pls., 3 text-figs., 1 table.

Frey, R. W., 1970, Trace fossils of Fort Hays Limestone Member of Niobrara Chalk (Upper Cretaceous), west-central Kansas: Kansas Univ. Paleont. Contr., art. 53 (Cretaceous 2), 41 p., app., 10 pls., 5 text-figs., 4 tables.

Gandolfi, R., 1942, Ricerche micropaleontologiche e stratigrapfiche sulla Scaglia e sul flysch Cretacici dei Dintorni di Balerna (Canton Ticino): Riv. Italiana Paleontologia e Stratigrafia, v. 48, no. 4, p. 1-160, Pls. 1-14, Text-Figs. 1-49.

——1955, The genus *Globotruncana* in northeastern Columbia: Bulls. Am. Paleontology, v. 36, p. 1-118, 10 pls., 12 text-figs.

Geinitz, H. B., 1872-1875, Das Elbthalgebirge in Sachsen, II: Palaeontographica, v. 20, pt. 2, p. 46-49, Pl. 12.

——1873, Ueber Inoceramen der Kreide Formation: Neues Jahrb. Mineralogie, Geologie, u. Paläontologie, p. 7-22.

Gilbert, G. K., 1896, The underground water of the Arkansas Valley in eastern Colorado: U.S. Geol. Survey, 17th Ann. Rept., 1895-1896, pt. 2, p. 551-601, Pl. 56-58, Text-Figs. 45-49.

Gill, J. R., and Cobban, W. A., 1966, The Red Bird section of the Upper Cretaceous Pierre Shale in Wyoming: U.S. Geol. Survey Prof. Paper 393-A, 73 p., 12 pls., 17 text-figs.

Goldfuss, A., 1836, Petrefacta Germaniae: Düsseldorf, Arnz and Co., p. 1-312, Pls. 1-90.

Gould, C. N., 1925, Index to the stratigraphy of Oklahoma: Oklahoma Geol. Survey Bull. 35, 115 p., with chart.

——1929, The Benton Cretaceous in Oklahoma: Oklahoma Acad. Sci. Proc., v. 8, p. 144-146.

Grabau, A. W., and Shimer, H. W., 1909, North American index fossils: Vol. 1, Invertebrates: New York, A. G. Seiler and Co., 443 p., illus.

Gryc, George, 1951 (1952), Paleontology, *in* Payne, T. G., and others, Geology of the Arctic slope of Alaska: U.S. Geol. Survey Oil and Gas Inv. Map OM-126, sheet 3, fig. 20, scale 1:1,000,000.

Haller, Wolfgang, 1963, Zur Makrofauna der Oberkreidesedimente im Gebiet von Spremberg-Weisswasser: Geol. Gesell. DDR Ber., v. 8, no. 2, p. 158.

Häntzschel, Walter, 1962, Trace fossils and Problematica, *in* Moore, R. C., ed., Treatise on invertebrate paleontology, Pt. W, Miscellanea: Boulder, Colo., Geol. Soc. America (and Univ. Kansas Press), p. W177–W245, Text-Figs. 109–149.

Hattin, D. E., 1962, Stratigraphy of the Carlile Shale (Upper Cretaceous) in Kansas: Kansas Geol. Survey Bull. 156, 155 p., 27 pls., 2 text-figs., 2 tables.

——1964, Cyclic sedimentation in the Colorado Group of west-central Kansas, *in* Merriman, D. F., ed., Symposium on cyclic sedimentation: Kansas Geol. Survey Bull. 169, v. 1, p. 205–217, 6 text-figs.

——1965a, Stratigraphy of the Graneros Shale (Upper Cretaceous) in central Kansas: Kansas Geol. Survey Bull. 178, 83 p., 5 pls., 15 text-figs., 8 tables.

——1965b, Upper Cretaceous paleontology and paleoecology of western Kansas: Geol. Soc. America, Ann. Mtg., Field. Conf. Guidebook, 69 p., 18 text-figs. (with section on Pierre Shale by W. A. Cobban).

——1967, Stratigraphic and paleoecologic significance of macroinvertebrate fossils in the Dakota Formation (Upper Cretaceous) of Kansas, *in* Teichert, C., and Yochelson, E. L., eds., Essays in Paleontology and stratigraphy (Raymond C. Moore volume): Kansas Univ. Dept. Geology Spec. Pub. 2, p. 570–589, 7 text-figs.

——1968, *Plesiacanthoceras wyomingense* (Reagan) from Graneros Shale and Greenhorn Limestone (Upper Cretaceous) of central Kansas: Jour. Paleontology, v. 42, p. 1084–1090, 4 text-figs.

——1971, Widespread, synchronously deposited, burrow-mottled limestone beds in Greenhorn Limestone (Upper Cretaceous) of Kansas and southeastern Colorado: Am. Assoc. Petroleum Geologists Bull., v. 55, p. 412–431, 23 figs.

——1975, Stratigraphy and depositional environment of Greenhorn Limestone (Upper Cretaceous) of Kansas: Kansas Geol. Survey Bull. 209, 128 p., 9 pls., 23 text-figs., 3 tables.

Hattin, D. E., and Frey, R. W., 1969, Facies relations of *Crossopodia* sp., a trace fossil from the Upper Cretaceous of Kansas, Iowa, and Oklahoma: Jour. Paleontology, v. 43, p. 1435–1440, 2 text-figs.

Heine, E., 1929, Die Inoceramen des Mittlewestfälischen Emschers und Unteren Senons: Preuss. Geol. Landesanst. Abh., N. F., no. 120, p. 1–124, Pls. 1–20.

Heinz, Rudolf, 1928, Über die Oberkreide—Inoceramen Südamerikas und ihre Beziehungen zu denen Europas und anderer Gebiete; Pt. 5: Mitt.-Miner.-Geol. Staatsinstitut Hamburg, no. 10, p. 42–97, Pls. 1–3.

——1932, Aus der neuen Systematik der Inoceramen (Inoceramen XIV): Mitt.-Miner.-Geol. Staatsinstitut Hamburg, no. 13, p. 1–26.

——1933 (1934), Inoceramen von Madagaskar über ihre Bedeutung für die Kreide-Stratigraphie. (Inoceramen XII): Deutsch. Geol. Gesell. Zeitschr., v. 85, no. 4, p. 241–259, Pls. 16–22.

Hermes, J. J., 1969, Late Albian foraminifera from the Subbetic of southern Spain: Geologie en Mijnbouw, v. 48, no. 1, p. 35–65, Pls. 1–8.

Hill, R. T., 1887, The Texas section of the American Cretaceous: Am. Jour. Sci., ser. 3, v. 34, p. 287–309, 4 text-figs.

——1889, A preliminary annotated check list of Cretaceous invertebrate fossils of Texas, accompanied by a short description of the lithology and stratigraphy of the system: Texas Geol. Survey Bull. 4, 57 p.

——1901, Geography and geology of the Black and Grand Prairies, Texas: U.S. Geol. Survey, 21st Ann. Rept., pt. 7, p. 5–666, 71 pls.

Hyatt, Alpheus, 1903, Pseudoceratites of the Cretaceous: U.S. Geol. Survey Mon. 44, 351 p., 47 pls.

Irwin, M. L., 1965, General theory of epeiric clearwater sedimentation: Am. Assoc. Petroleum Geologists Bull., v. 49, p. 445–459, 12 text-figs.

Jeannet, A., 1922, Description d'une espèce nouvelle d'inocérame: *in* Heim, Arnold, and Jeannet, Alphonse, Crétacique supérieur à Inocérames et Éocène de la Nouvelle-Calédonie: Soc. Géol. France Bull., v. 22, p. 251–253, Text-Fig. 5.

Johnson, D. W., 1903, The geology of the Cerillos Hills, New Mexico: New Mexico School Mines Quart., 1903-1904, v. 24, p. 204 (reprinted by Press of the New Era Printing Co., Lancaster, Pa., 1904).

Jones, D. L., and Gryc, George, 1960, Upper Cretaceous pelecypods of the genus *Inoceramus* from northern Alaska: U.S. Geol. Survey Prof. Paper 334-E, p. 149-165, Pls. 15-23, 4 text-figs., 2 tables.

Kauffman, E. G., 1961, Mesozoic paleontology and stratigraphy, Huerfano Park, Colorado; Vols. 1 and 2 [Ph.D thesis]: Univ. Michigan, 1467 p., 67 pls., 62 text-figs., 1 chart *and* Dissert. Abs., 1962, v. 22, no. 8, p. 2754-2755.

——1965, Middle and Late Turonian oysters of the *Lopha lugubris* group: Smithsonian Misc. Colln., v. 148, no. 6, Publ. 4602, 93 p., 8 pls., 18 text-figs, 5 tables.

——1966a, Notes on the Cretaceous Inoceramidae (Bivalvia) of Jamaica: Jamaica Geol. Soc. Jour., v. 8, p. 32-40, 1 table, June 1967.

——1966b, Taxonomic, ecologic, and evolutionary significance of interior shell morphology in the Inoceramidae (Mesozoic Bivalvia) [abs.]: Geol. Soc. America Spec. Paper 87, p. 86.

——1967a, Coloradoan macroinvertebrate assemblages, central Western Interior, United States, *in* Kauffman, E. G., and Kent, H. C., eds., Paleoenvironments of the Cretaceous seaway, a symposium: Colorado School Mines Spec. Pub., preprints of papers, p. 67-143, 12 text-figs.

——1967b, Cretaceous *Thyasira* from the Western Interior of North America: Smithsonian Misc. Colln., v. 152, no. 1, pub. 4695, 159 p., 5 pls., 18 text-figs., 7 tables.

——1969a, Cretaceous marine cycles of the Western Interior: Mountain Geologist, v. 6, p. 227-245, 4 text-figs.

——1969b, Form, function and evolution, *in* Moore, R. C., ed., Treatise on invertebrate paleontology, Pt. N, Vol. 1, Mollusca 6, Bivalvia: Boulder, Colo., Geol. Soc. America (and Univ. Kansas Press), p. N129-N205, 17 text-figs.

——1970, Population systematics, radiometrics, and zonation—A new biostratigraphy: North America Paleont. Convention, Proc., Pt. F, p. 612-666, 10 text-figs.

——1973, Cretaceous Bivalvia, *in* Hallam, A., ed., Atlas of paleobiogeography: Amsterdam, Elsevier Pub. Co., p. 353-383, 10 text-figs.

——1977, Cretaceous geochronology for the Western Interior United States: New data and applications: Geology, v. 5 (in press).

Kauffman, E. G., and Kent, H. C., 1969, Cretaceous biostratigraphy of Western Interior United States [abs.]: Geol. Soc. America, Abs. for 1968, Spec. Paper 121, p. 156.

Kauffman, E. G., Powell, J. D., and Hattin, D. E., 1969, Cenomanian-Turonian facies across the Raton Basin: Mountain Geologist, v. 6, p. 93-118, 4 text-figs.

Kayser, F.H.E., 1893, Lehrbuch der geologischen Formationskunde: Lehrbuch der Geologie, v. 1, 2, Stuttgart, F. Enke, 532 p., illus., table, maps.

——1924, Text book of comparative geology (English edition translated by Philip Lake): London, S. Sonnenschein & Co.; New York, Macmillan & Co., xii + 426 p., illus.

Kennedy, W. J., 1971, Cenomanian ammonites from southern England: Spec. Papers Palaeontology, no. 8, 133 p., 64 pls.

Kent, H. C., 1967, Microfossils from the Niobrara Formation (Cretaceous) and equivalent strata in northern and western Colorado: Jour. Paleontology, v. 41, p. 1433-1456, Pls. 183, 184, 8 text-figs.

——1968, Biostratigraphy of Niobrara-equivalent part of Mancos Shale (Cretaceous) in northwestern Colorado: Am. Assoc. Petroleum Geologists Bull., v. 52, pt. 1, p. 2098-2115, 10 text-figs.

Loeblich, A. R., and Tappan, Helen, 1961, Cretaceous planktonic foraminifera, Pt. 1—Cenomanian: Micropaleontology, v. 7, p. 257-304, Pls. 1-8.

Logan, W. N., 1898, The invertebrates of the Benton, Niobrara and Fort Pierre Groups: Kansas Geol. Survey, v. 4, p. 431-518, illus.

——1899, Some additions to Cretaceous invertebrates of Kansas: Kansas Univ. Quart., v. 8, ser. A, p. 87-98, illus.

MacKenzie, D. B., 1971, Post-Lytle Dakota Group on west flank of Denver Basin, Colorado: Mountain Geologist, v. 8, p. 91-131, 4 pls., 19 text-figs., 3 tables.

Mantell, G. A., 1822, The fossils of South Downs, or illustrations of the geology of Sussex: London, Lupton Relfe, xiv + 327 p., 42 pls., 1 geol. map, text illus.

Maury, C. J., 1936, O cretaceo de Sergipe: Brazil Serv. Geológico e Mineralógico Mon. 11, p. 1-283, Pls. 1-28.

Maxwell, R. A., Lonsdale, J. T., Hazzard, R. T., and Wilson, J. A., 1967, Geology of Big Bend National Park, Brewster County, Texas: Texas Univ. Pub. No. 6711, 320 p., 11 pls., 152 text-figs., 16 tables.

McAlester, A. L., 1962, Mode of preservation in early Paleozoic pelecypods and its morphologic and ecologic significance: Jour. Paleontology, v. 36, p. 69-73, Pl. 16, 1 text-fig.

McLaughlin, T. G., 1954, Geology and ground-water resources of Baca County, Colorado: U.S. Geol. Survey Water-Supply Paper 1256, 232 p., 2 pls., 54 text-figs., 9 tables.

McLearn, F. H., 1926, New species from the Coloradoan of Lower Smoky and Lower Peace Rivers, Alberta, Canada: Canada Geol. Survey Bull. 42, p. 117-126, 4 pls.

——1945, Revision of the Lower Cretaceous of Western Interior of Canada (2nd ed.): Canada Geol. Survey Paper 44-17, p. 1-14, Pls. 1-12.

Meek, F. B., 1872, Preliminary list of the fossils collected in Utah and Wyoming territories, with descriptions of a few new species: U.S. Geol. Survey Montana (Hayden), Ann. Rept., no. 5, p. 373-377.

——1876, *in* Meek, F. B., and Hayden, F. V., Report on the invertebrate Cretaceous and Tertiary fossils of the Upper Missouri Country: Rept. U.S. Geol. Survey Territories, v. IX, 629 p., 45 pls.

Meek, F. B., and Hayden, F. V., 1857, Descriptions of new species and genera of fossils, collected by Dr. F. V. Hayden in Nebraska Territory; with some remarks on the Tertiary and Cretaceous formations of the northwest: Acad. Nat. Sci. Philadelphia, Proc. 1857, p. 117-148.

Mello, J. F., 1969, Foraminifera and stratigraphy of the upper part of the Pierre Shale and lower part of the Fox Hills Sandstone (Cretaceous), north-central South Dakota: U.S. Geol. Survey Prof. Paper 611, 121 p., 11 pls., 14 text-figs.

——1971, Foraminifera from the Pierre Shale (Upper Cretaceous) at Red Bird, Wyoming: U.S. Geol. Survey Prof. Paper 393-C, 54 p., 7 pls., 4 text-figs.

Moreman, W. L., 1927, Fossil zones of the Eagle Ford of north Texas: Jour. Paleontology, v. 1, p. 89-101.

——1933; Stratigraphy *in* Adkins, W. S., The Mesozoic systems in Texas; Vol. 1: Texas Univ. Bull. 3232, p. 424-426 (1932).

——1942, Paleontology of the Eagle Ford Group of north and central Texas: Jour. Paleontology, v. 16, p. 192-220, 4 pls., 2 text-figs.

Morris, John, 1854, A catalogue of British fossils comprising genera and species hitherto described (2nd ed.): London, pub. by author, vii + 372 p.

Morrow, A. L., 1934, Foraminifera and Ostracoda from the Upper Cretaceous of Kansas: Jour. Paleontology, v. 8, p. 186-205, Pls. 29-31.

Nagao, T., and Matsumoto [originally Matumoto], T., 1939, A monograph of the Cretaceous *Inoceramus* of Japan; Pt. 1; Hokkaido Imp. Univ. Fac. Sci. Jour., ser. 4 (Geol. Mineral.), v. 4, p. 241-299, Pls. 1-12.

Newell, N. D., and Hertlein, L. G., 1969, Family Entoliidae Korobkov, 1960, *in* Moore, R. C., ed., Treatise on invertebrate paleontology, Pt. N, Vol. 1, Mollusca 6, Bivalvia: Boulder, Colo., Geol. Soc. America (and Univ. Kansas Press), p. N346-N347, Fig. C71.

Obradovich, J. D., and Cobban, W. A., 1973, A time-scale for the Late Cretaceous of the Western Interior of North America: Geol. Assoc. Canada Ann. Meeting, Program and Abs., May 23-26, 1973, p. 42-43.

——1975, A time scale for the Late Cretaceous of the Western Interior of North America, *in* Caldwell, W.G.E., ed., The Cretaceous System in the Western Interior of North America: Geol. Assoc. Canada Spec. Paper No. 13, p. 31-54.

d'Orbigny, A. de, 1843, Paléontologie Française; Description des mollusques et rayonnés fossiles Terrains cretaces; Vol. 3, Lamellibranches: Paris, G. Masson, 807 p., Pls. 237-489.

Pergament, M. A., 1966, Zonal stratigraphy and *Inoceramus* of the lowermost Upper Cretaceous on the Pacific Coast of the USSR: Acad. Sci. USSR, Geol. Inst., Trans., v. 146, 83 p., 36 pls.

Pessagno, E. A., Jr., 1967, Upper Cretaceous planktonic Foraminifera from the western Gulf coastal plain: Palaeontographica Americana, v. 5, no. 37, p. 245–445, Pls. 48–101, 65 text-figs.
——1969, Upper Cretaceous stratigraphy of the western Gulf Coast area, Mexico, Texas and Arkansas: Geol. Soc. America; Mem. 111, 139, p., 60 pls.
Pictet, F. J., 1868–1871, Matériaux pour la paléontologie Suisse sur les fossiles du Jura et des Alpes: Genève, H Georg, 5th ser., 197 p., 28 pls.
Plummer, N. V., and Romary, J. F., 1942, Stratigraphy of the pre-Greenhorn Cretaceous beds of Kansas: Kansas Geol. Survey Bull. 41, pt. 9, p. 313–348, 2 pls., 4 text-figs., 2 tables.
Powell, J. D., 1963a, Cenomanian-Turonian (Cretaceous) ammonites from Trans-Pecos Texas and northeastern Chihuahua, Mexico: Jour. Paleontology, v. 37, p. 309-322, Pls. 31-34, 3 text-figs.
——1963b, Turonian (Cretaceous) ammonites from northeastern Chihuahua, Mexico: Jour. Paleontology, v. 37, p. 1217–1232, Pls. 166–171, 6 text-figs.
——1965, Late Cretaceous platform-basin facies, northern Mexico and adjacent Texas: Am. Assoc. Petroleum Geologists Bull., v. 49, p. 511–525, 8 text-figs.
——1968, Woodbine-Eagle Ford transition, Tarrant Member; *in* Dodge, C. F., ed., Stratigraphy of the Woodbine Formation, Tarrant County, Texas, *in* Field Trip Guidebook, 2nd Ann. Mtg., South-Central Sec., Geol. Soc. America: p. 27–43, 2 text-figs.
——ed., 1970, Field trip guidebook for the first interamerican micropaleontological colloquium, July 19–30: Dallas, Univ. Texas at Dallas (published privately).
——1971, Relationship between the Late Cretaceous Western Interior and Southern seaways in the United States: Geol. Soc. America Abs. with Programs, v. 3, no. 7, p. 674–675.
Pusch, G. G., 1837, Polens Paläontologie; oder Abbildung und Beschreibung der vorzüglichsten und der noch unbeschriebenen Petrefakten aus den Gebirgsformationen in Polen, Volhynien und den Karpathen nebst einigen allgemeinen Beiträgen zur Petrefaktenkunde und einem Versuch zur Vervollständigung der Geschichte des europäischen Auer-Ochsen: Stuttgart, p. 1–218, pls. 1–16.
Reagan, A. B., 1927, Fossils from the Mancos-Dakota-Tununk Formations in the vicinity of Steamboat, Arizona: Indiana Acad. Sci. Proc., v. 36 (1926), p. 119–127, Text-Figs. 10–12.
Reeside, J. B., Jr., 1923, The fauna of the so-called Dakota Formation of northern central Colorado and its equivalent in southeastern Wyoming: U.S. Geol. Survey Prof. Paper 131-H, p. 199–207, Pls. 45–50.
——1957, Paleoecology of Cretaceous seas of the Western Interior of the United States, *in* Ladd, H.S., ed., Paleoecology, Vol. 2, *of* Treatise on marine ecology and paleoecology: Geol. Soc. America Mem. 67, v. 2, p. 505–541.
Reeside, J. B., Jr., and Cobban, W. A., 1960, Studies of the Mowry Shale (Cretaceous) and contemporary formations in the United States and Canada: U.S. Geol. Survey Prof. Paper 355, 126 p., 58 pls., 30 text-figs.
Rhoads, D. C., 1970, Mass properties, stability, and ecology of marine muds related to burrowing activity, *in* Crimes, T. P., ed., Trace fossils: Geol. Jour. Spec. Issue 3, p. 391–406.
Rossi Ronchetti, Carla, 1961, Fossili cretacei di Pull-i-Khumri (Afghanistan): Riv. Italiana Paleontologia e Stratigrafia, v. 67, no. 4, p. 341–368, Pl. 3.
Rothrock, E. P., 1925, Geology of Cimarron County, Oklahoma: Oklahoma Geol. Survey Bull. 34, p. 7–92, 24 pls., 3 text-figs.
Rutsch, R. F., and Salvador, Amos, 1954, Mollusks from the Cogollo and La Luna Formations (Cretaceous) of the Chejende area, western Venezuela: Jour. Paleontology, v. 28, p. 417–426, Pl. 40, Text-Figs. 1, 2.
Schiel, James, 1855, List and description of organic remains collected during the exploration of the central Pacific railroad line 1853–1854, *in* Beckwith, E. G., Report of explorations for the Pacific Railroad on the line of the forty-first parallel of north latitude: U.S. Pacific R.R. Expl. (U.S. 33rd Cong., 1st sess., House Executive Doc. 129, v. 18, pt. 2), p. 134–135 (and U.S. 33rd Cong., 2nd sess., Senate Executive Doc. 78 and House Executive Doc. 91, pt. 2, p. 108–109, Pls. 1–3).

Schlotheim, E. T., 1813, Beiträge zur Naturgeschichte der Versteinerungen in geognotischer Hinsicht: Leonhard's Taschenbuch für Mineralogie, v. 7, p. 93.

Schluter, C., 1877, Kreidebivalven zur Gattung *Inoceramus:* Paleontographica, v. 24, p. 251–288, Pls. 1–4.

Schoff, S. L., 1943, Geology and groundwater resources of Cimarron County, Oklahoma, with a section on Mesozoic stratigraphy by John Willis Stovall: Oklahoma Geol. Survey Bull. 64, 317 p., 23 pls., 27 text-figs., 24 tables.

Scott, G. R., and Cobban, W. A., 1964, Stratigraphy of the Niobrara Formation at Pueblo, Colorado: U.S. Geol. Survey Prof. Paper 454L, p. L1–L30, Pls. 1–11, 3 tables, 9 text-figs.

Scott, R. W., 1970, Stratigraphy and sedimentary environments of Lower Cretaceous rocks, southern Western Interior: Am. Assoc. Petroleum Geologists Bull., v. 54, p. 1225–1244, 5 text-figs., 1 table.

Seilacher, Adolf, 1969, Paleoecology of boring barnacles: Am. Zoologist, v. 9, p. 705–719, 8 figs.

Seitz, Otto, 1934 (1935), Die Variabilität des *Inoceramus labiatus* Schlotheim: Preuss. Geol. Landesanst. Jahrb., v. 55, p. 429–474, Pls. 1–5, 9 text-figs.

Shaw, A. B., 1964, Time in stratigraphy: New York, McGraw-Hill, 365 p., illus.

Shimer, H. W., and Blodgett, M. E., 1908, The stratigraphy of the Mt. Taylor Region, New Mexico: Am. Jour. Sci., 4th ser., v. 25, p. 53–67.

Shimer, H. W., and Shrock, R. R., 1955, Index fossils of North America: New York, John Wiley & Sons, i–ix + 837 p., 303 pls.

Shumard, B. F., 1860, Observations upon the Cretaceous strata of Texas: St. Louis Acad. Sci. Trans., v. 1, p. 582–590.

Siemers, C. T., 1971, Stratigraphy, paleoecology, and environmental analysis of upper part of Dakota Formation (Cretaceous), central Kansas [Ph.D. thesis]: Bloomington, Indiana Univ., 287 p., 23 pls., 38 text-figs., 13 tables.

Simionescu, I., 1899, Fauna cretacica superiora de la Urmös (Transilvania): Burcuresci, Acad. Romana, Publicatiunile fondu vi vasilie adamachi, v. 1, no. 4, p. 39, Pl. 3.

Sohl, N. F., 1960, Archaeogastropoda, Mesogastropoda and stratigraphy of the Ripley, Owl Creek, and Prairie Bluff Formations: U.S. Geol. Survey Prof. Paper 331-A, 151 p., pls. 1–18, 11 text-figs., 1 table.

——1967, Upper Cretaceous gastropod assemblages of the Western Interior of the United States, *in* Kauffman, E. G., and Kent, H. C., eds., Paleoenvironments of the Cretaceous seaway—a symposium: Golden, Colorado School of Mines Spec. Pub., 37 p., 9 text-figs.

Sornay, Jacques, 1965, Le faune d'inocérames du cénomanien et du turonien inférieur du sud-ouest de Madagascar: Annales Paléontologie, Invertebres, v. 51, p. 1–18, Fig. 1, illus., maps.

——1966, Idées actuelles sur les inocérames d'après divers travaux récents: Annales Paléontologie, Invertebres, v. 52, pt. 1, p. 59–92, Text-Figs. 1–10.

Sowerby, J. de C., 1825, The mineral conchology of Great Britain: London, Richard Taylor, v. 5, p. 1–168.

——1829, The mineral conchology of Great Britain: London, Richard Taylor, v. 6, p. 1–250.

Speden, I. G., 1967, Revision of *Syncyclonema* (Upper Cretaceous) and comparison with other small pectinid bivalves and *Entolium:* Postilla, no. 110, p. 1–36, Pls. 1–7.

Stanton, T. W., 1893 (1894), The Colorado Formation and its invertebrate fauna: U.S. Geol. Survey Bull. 106, 288 p., 45 pls.

Stenzel, H. B., 1971, Oysters, *in* Moore, R. C., and others, eds., Treatise on invertebrate paleontology, Pt. N, Vol. 3, Mollusca 6, Bivalvia: Boulder, Colo., Geol. Soc. America (and Univ. Kansas Press), p. N953–N1224, Text-figs. J1–J153.

Stephenson, L. W., 1941, The larger invertebrate fossils of the Navarro Group of Texas: Texas Univ. Pub. 4101, 641 p., 95 pls., 13 text-figs., 6 tables.

——1952, Larger invertebrate fossils of the Woodbine Formation (Cenomanian) of Texas: U.S. Geol. Survey Prof. Paper 242, 226 p., 59 pls., 8 text-figs.

——1955, Basal Eagle Ford fauna (Cenomanian) in Johnson and Tarrant Counties Texas: U.S. Geol. Survey Prof. Paper 274-C, p. 53–67, Pls. 4–7.

Stephenson, L. W., and Reeside, J. B., Jr., 1938, Comparisons of Upper Cretaceous deposits of Gulf region and Western Interior region: Am. Assoc. Petroleum Geologists Bull., v. 22, p. 1629-1638, 3 text-figs.

Stephenson, L. W., Kind, P. B., Monroe, W. H., and Imlay, R. W., 1942, Correlation of the outcropping Cretaceous rocks of the Atlantic and Gulf Coastal Plain and Trans-Pecos region: Geol. Soc. America Bull., v. 53, p. 435-448, with chart.

Stoliczka, F., 1869 (1871), Cretaceous fauna of southern India: Paleont. Indica, Vol. 3, Ser. 6, The Pelecypoda, with a review of all known genera of this class, fossil and Recent: Calcutta, Geol. Survey India Mem., 537 p., 50 pls.

Stovall, J. W., 1938, The Morrison of Oklahoma and its dinosaurs: Jour. Geology, v. 46, p. 583-600, 3 text-figs., 5 tables.

——1942, The Mesozoic of western Oklahoma [abs.]: Compass, v. 22, p. 327.

——1943, Stratigraphy of the Cimarron Valley (Mesozoic rocks), *in* Schoff, S. L., Geology and ground water resources of Cimarron County, Oklahoma: Oklahoma Geol. Survey Bull. 64, p. 43-100, 11 pls., 1 text-fig., 6 tables.

Strombeck, A. V., 1863, Ueber die Kreide am Zeltberg bei Lüneburg: Deutsch. Geol. Gesell. Zeitschr., v. 15, p. 97-187, Pl. 4.

Tappan, Helen, 1940, Foraminifera from the Grayson Formation of northern Texas: Jour. Paleontology, v. 14, p. 93-126, Pls. 14-19.

——1962, Foraminifera of the Arctic Slope of Alaska, Pt. 3, Cretaceous Foraminifera: U.S. Geol. Survey Prof. Paper 236-C, 209 p., Pls. 29-58, 18 text-figs.

Thomel, Gérard, 1969, Réflexions sur les genres *Eucalycoceras* et *Protacanthoceras* (Ammonoidea): Acad. Sci. Comptes Rendus, v. 268, ser. D, Nat. Sci., p. 649-652.

Thorson, Gunnar, 1957, Bottom communities (sublittoral or shallow shelf); *in* Hedgpeth, J. W., ed., Ecology, Vol. 1 *of* Treatise on marine ecology and paleoecology: Geol. Soc. America Mem. 67, p. 461-534, 20 text-figs.

Trexler, D. W., 1967, Stratigraphic distribution of Upper Cretaceous nannoplankton (coccoliths) in central and northern Colorado and the Black Hills region: Jour. Paleontology, v. 41, p. 1355-1364, 4 text-figs.

Tröger, K. A., 1967, Zur Paläontologie, Biostratigraphie, and faziellen Ausbildung der unteren Oberkreide (Cenoman bis Turon); Pt. 1, Paläontologie und Biostratigraphie der Inoceramen des Cenomans bis Turons Mitteleuropas: Dresden, Statatl. Mus. Mineral. Geol., Abh., v. 12, p. 13-207, 14 pls.

Tsagareli, A. L., 1963, Upper Cretaceous fauna from Dagestan: Akad. Nauk Grunzin. SSR Inst. Geologicheskii, v. 13, no. 18, p. 79-108, Pls. 1-7.

Veatch, A. C., 1907, Geography and geology of a portion of southwestern Wyoming, with special reference to coal and oil: U.S. Geol. Survey Prof. Paper 56, 178 p., 26 pls., 9 text-figs.

Waage, K. W., 1953, Refractory clay deposits of south-central Colorado: U.S. Geol. Survey Bull. 993, 104 p., 17 text-figs., 7 tables.

——1955, Dakota Group in northern Front Range foothills, Colorado: U.S. Geol. Survey Prof. paper 274-B, p. 15-51, 19 text-figs.

——1961, Stratigraphy and refractory clayrocks of the Dakota Group along the Northern Front Range, Colorado: U.S. Geol. Survey Bull. 1102, 154 p., 13 text-figs., 2 tables.

Wall, J. H., 1967, Cretaceous Foraminifera of the Rocky Mountain Foothills, Alberta: Research Council Alberta Bull. 20, 119 p., 19 pls.

Warren, P. S., 1930, Description of new species of fossils from parts of Peace River and Grand Prairie Districts, Alberta: Research Council Alberta Rept. 21, p. 57-68, Pls. 3-7.

Weimer, R. J., 1960, Upper Cretaceous stratigraphy, Rocky Mountain area: Am. Assoc. Petroleum Geologists Bull., v. 44, p. 1-20, 10 text-figs, 1 table.

White, C. A., 1876 (1877), Report upon the invertebrate fossils collected in portions of Nevada, Utah, Colorado, New Mexico, and Arizona, by parties of expeditions of 1871, 1872, 1873 and 1874: U.S. Geog. Survey Terr. West of 100th Meridian, Rept., v. 4, Paleontology, pt. 1, 219 p., 21 pls.

Williams, G. D., and Baadsgaard, H., 1973, Potassium-argon dates and Cretaceous biostratigraphy in eastern Saskatchewan: Geol. Assoc. Canada Ann. Mtg. Program and Abs., May 23-26, 1973, p. 50.

——1975, Potassium-argon dates and Upper Cretaceous biostratigraphy in eastern Saskatchewan, *in* Caldwell, W.G.E., ed., The Cretaceous System in the Western Interior of North America: Geol. Assoc. Canada Spec. Paper 13, p. 417–426.

Williams-Mitchell, E., 1948, The zonal value of foraminifera in the Chalk of England: Geologists' Assoc. Proc., v. 59, pt. 2, p. 91–112, Pls. 8–10.

Woods, Henry, 1911, Monograph of the Cretaceous Lamellibranchia of England, Vol. 2, Pt. 7: Palaeontographical Soc., v. 64 (1910), p. 261–284, 6 pls., 12 text-figs.

Wright, C. W., 1957, Mesozoic Ammonoidea, *in* Arkell, W. J., Kummel, B., and Wright, C. W., Treatise on invertebrate paleontology, Pt. L, Mollusca 4, Cephalopoda, Ammonoidea (R. C. Moore, ed.): Boulder, Colo., Geol. Soc. America (and Univ. Kansas Press), p. L80–L437.

——1963, Cretaceous ammonities from Bathurst Island, Northern Australia: Palaeontology, v. 6, p. 597–614.

Young, Keith, 1963, Upper Cretaceous ammonites from the Gulf Coast of the United States: Texas Univ. Pub. 6304, Bur. Econ. Geology, 373 p., 81 pls., 34 text-figs., 13 tables.

MANUSCRIPT RECEIVED BY THE SOCIETY JANUARY 23, 1975
REVISED MANUSCRIPT RECEIVED MAY 20, 1976
MANUSCRIPT ACCEPTED JUNE 11, 1976